American Institute of Chemical Engineers

Topical Conference on Distillation 2006

Held at the 2006 AIChE Spring National Meeting

April 23-27, 2006
Orlando, Florida, USA

Printed from e-media with permission by:

Curran Associates, Inc.
57 Morehouse Lane
Red Hook, NY 12571
www.proceedings.com

ISBN: 978-1-60423-527-2

Some format issues inherent in the e-media version may also appear in this print version.

American Institute of Chemical Engineers

Topical Conference on Distillation 2006

Held at the 2006 AIChE Spring National Meeting

AIChE

April 23-27, 2006
Orlando, Florida, USA

American Institute of Chemical Engineers

Topical Conference on Distillation
2006

TABLE OF CONTENTS

Session 30 - Separations Research Programs

This session will showcase the many academic and industrial research programs dedicated to improving separations and associated processes.

CoChair: Zarko Olujic

Chair: Nicholas F. Urbanski

Session 58 - Distillation Honors: John Farone

This session will be dedicated to showcasing the lifetime of engineering experience of our honoree. While primarily an invite-only session, authors may submit topics related to John Farones and his work for consideration.

CoChair: Zarko Olujic

Chair: Nicholas F. Urbanski

Mark W. Pilling

15 Evaluation of Structured Packing Hydraulic Performance Using X-Ray Computed Tomography
Christian W. Green, R. Bruce Eldridge

16 Pressure Drop of Packed Column Internals
Armin Rix, Zarko Olujic

17 A Deep-Knowledge Based Evolutionary Operation Approach for Distillation Process Optimization
Simon Xu, **Lowell Pless**

26 Extractive Distillation: a Process Improvement Project
James P. Farone, Menno M. Kroes, Mark Mueller

Session 88 - Advances in Distillation Equipment & Applications I

This session will address the current research and application of new tray and packing technologies.

CoChair: Izak Nieuwoudt

Chair: Waldo E. De Villiers

33 Dividing Wall Column Revamp Optimizes Xylene Production
Bernie Slade, **Berne K. Stober**, Dave Simpson

39 Investigation on a Novel Double-Effective Packing Element
Friederike Kutzer, Johann Stichlmair

48 Measurement of Packed Column Gas/Liquid Contact Areas for CO2 Absorption
Frank Seibert, J. Christopher Lewis

49 Debottlenecking an Acn Heads & Dry Column with Vg Af (V-Grid Anti-Fouling) Trays
Giuseppe Mosca, Elena Tacchini, Joshi Chandrakant, Shubhangi Jaguste, Prabhakar Nag, Kanubhai Kalidas Parmar

50 Modeling the Minimum Entrainment on Sieve Trays
Shuzo Ohe

Session 129 - Advances in Distillation Equipment & Applications II

This session will address the current research and application of new tray and packing technologies.

CoChair: Waldo E. De Villiers

Chair: Izak Nieuwoudt

70 Testing of Mellapakplus M452.Y at Fractionation Research, Inc
Mark W. Pilling, Florian Kehrer, Lothar Spiegel

71 Foaming Effect on Random Packing Performance
G.X. Chen, Tony Cai, **Karl T. Chuang**, A Afacan

72 Blended Packed Bed Performance

SuperBlend 2-Pac™ Packing
Frank Rukovena Jr., Bob Lobaugh

97 How to Surpass Conventional and High Capacity Structured Packings with Raschig Super-Pak
Michael Schultes, Simon Chambers

110 Radioisotope Technology - Benefits & Limitations in Packed Bed Tower Diagnostics
Scott Vidrine

Session 176 - Advances in Distillation Modeling & Processes

This session will address current research and application of new process modeling techniques.

Chair: Ross Taylor

112 Tray Design Techniques at Low Liquid Load Conditions
Daniel R. Summers, Andrew W. Sloley

113 Distillation Column Flooding Predictor
M. Geoff Mphahlele, R. Bruce Eldridge

115 Computer-Aided Graphical Tool for Complex Column Design
Michaela Vrey, Simon Thornhill Holland, Diane Hildebrandt, **David Glasser**

117 Heat-Integrated Distillation Sequence Synthesis
Santosh Jain, Jin-Kuk Kim, **Robin Smith**

118 Synthesis and Design of Energy Efficient Distillation
Angelo Lucia, Ross Taylor, Amit Amale

120 How to Decide When and How Much to Use Reactive Distillation: a Case Study on Mtbe System
Jean Lubilanji, David Glasser, Diane Hildebrandt

30a Recent Distillation Research at FRI

G.X. Chen and Tony Cai. Fractionation Research, Inc., P.O. Box 2108, Stillwater, OK 74076

Fractionation Research, Inc. (FRI) has carried out industrial scale distillation research with hydrocarbon systems at column pressures from 10 mmHg to 34 bar in its 1.22 and 2.44 m columns since 1952. It determined and studied the hydraulic and mass transfer performances of various trays and packings and provided complete design manuals to its memberships and industry. Recent FRI fundamental distillation research includes foaming effect on tray and packing performances, wetting effect on packing performance, high viscosity effect on tray hydraulics, and random packing performance under high pressure conditions. Recent FRI distillation equipment research studied the performances of new structured packings and new conventional and non-conventional trays. The most significant results will be discussed and summarized.

30b The University of Texas Separations Research Program

R. Bruce Eldridge, Chemical Engineering, University of Texas at Austin, 1 University Station C0400, CPE 4.408, Austin, TX 78712 and Frank Seibert, Separations Research Program, University of Texas at Austin, 10100 Burnet Road, Bldg 133/CEER, Austin, TX 78758.
The presentation will address on-going and planned separations activities at the University of Texas and will reflect on the Separation Research Program's twenty year history of providing service to the chemical and refining industry.

30c Modeling Reboilers with HTRI Technology

R. Stanley Kistler, HTRI, 150 Venture Drive, College Station, TX 77845

Text Not Available

30d Optimizing the Performance of High Capacity Corrugated Sheet Structured Packings

Zarko Olujic[1], Thomas Rietfort[2], Bjoern Kaibel[2], Helmut Jansen[2], and Egon Zich[2]. (1) Delft University of Technology, Leeghwaterstraat 44, Delft, 2628 CA, Netherlands, (2) Engineering, Julius Montz GmbH, Hofstrasse 82, Hilden, 40723, Germany
Experimental evaluation of the performance of both conventional and high capacity Montz packings indicated that with increasing specific area the installed area is used less efficiently. To alleviate this, most recently, the effects of changes in corrugated sheet geometry have been studied, both by model and by experiment. This provided a better insight into the relationship between the pressure drop and mass transfer efficiency. With optimized geometry the pressure drop was redistributed to increase efficiency without a penalty on the capacity side. The total reflux measurements were carried out at atmospheric pressure conditions with cyclohexane/n-heptane system using TU Delft facilities (0.45 m internal diameter column) as well as with chlorobenzene/ethylbenzene system using Bayer Technical Services facilities (0.6 m I.D.) column. The experimental results were used to validate the Delft model, which proved capable of predicting accordingly the studied effects and consequently can be recommended to be used as a conceptual design tool.

30e Optimum Product Yields of Vacuum Distillation Unit from Petroleum Refinery Process

Julio Pereira Maia[1], Edvaldo R. Morais[2], Igor R. S. Victorino[3], and **M. R. Wolf Maciel**[1]. (1) Chemical Engineering, UNICAMP, POBox 6066, Campinas, 13084759, Brazil, (2) Department of Chemical Processes, School of Chemical Engineering - Laboratory of Optimization, Design and Advanced Control, State University of Campinas (UNICAMP), Campinas, Brazil, (3) Department of Chemical Processes. School of Chemical Engineering - Laboratory of Optimization, Design and Advanced Control, State University of Campinas (UNICAMP), Campinas, Brazil

Vacuum distillation is employed on heavy petroleum fractions, with large applications on heavy and ultra-heavy crude processing, motivated by difficulties founded in processing these crudes at atmospheric pressure, due to the boiling range of the compounds in their compositions. These compounds would be easily decomposed if they were distillated under atmospheric pressure. The products obtained from this unit are used as feedstock for Fluid Catalytic Cracking Unit (FCCU). The relevance of this work can be based on many aspects, such as the trend of refineries in promoting modifications in vacuum towers and their related systems. Economically, due to the need of increasing the yield of gas oil production, used as feed in the Fluid Catalytic Cracking Unit (FCCU), and the production of asphalt and other products, as well as the need of a better recovery and use of the heavy fractions of the crude, caused by the raising market value of petroleum crudes. Technologically, due to the need of new technologies that could allow a higher participation of Brazilian crudes in national refineries processes, since the majority of crudes produced nowadays in Brazil are composed by heavy crudes, with a high amount of residue produced. The main objective of this work is the comprehension of the steady state and dynamic behavior of the system in order to define on-line optimization strategies and control. To achieve the objective a distillation vacuum column is established in a simulator with all auxiliary equipment, defining and characterizing feed and products, evaluating convergence criteria of the entire system in the steady state, followed by an optimization of the system. The dynamic behavior of the system is simulated, in order to evaluate the impact of load changes, even in operational variables as in design variables.

The steady state simulation allow to take into account the whole plant with the main auxiliary equipment, and provided the study of many conditions, as well as a guide to choose variables to be manipulated in order to achieve product requirements. Some blends of crudes were tested, and optimization in steady state were applied, according to energy consumption and product yield goals. The optimization in the steady state has lead the choose of a set of control variables, and respective set-points. The dynamic behavior of allowed to evaluate the impact of load changes, with an atmospheric column placed before the vacuum tower. The effect of heat load changes in furnace, as well as heat load changes in the condenser of vacuum column were considered, and the set of control variables reacted to the load changes, even in feed changes, such as feed flow rate and others.

Simulation of Ternary Distillation in a Heat Integrated Distillation Column (HIDiC) with a Rate-Based Model

Keigo Matsuda[1], Toshinari Nakanishi[2], Koichi Iwakabe[3], Kejin Huang[1], Takuji Yamamoto[1], Akira Endo[1], Takao Ohmori[1] and Masaru Nakaiwa[1,*]

(1)Research Institute for Innovation in Sustainable Chemistry, National Institute of Advanced Industrial Science and Technology, Central 5, 1-1-1, Higashi, Tsukuba, 305-8565, Japan;
[*]tel: +81-29-861-4661, e-mail: nakaiwa-m@aist.go.jp;
(2)Kimura Chemical Plants Co. Ltd., 2-1-2, Terajima Kuise, Amagasaki, Hyogo, 660-8567, Japan
(3)Tokyo Institute of Technology, 2-12-1, Ookayama, Meguro, Tokyo, 152-8552, Japan

A simulation technique has been developed to calculate HIDiC models. A process simulator "gPROMS" for prediction of separation performance of the HIDiC with structured packing is developed by rate-based model using Maxwell-Stefan equations. Comparison of observed data from bench-scale (benzene-toluene system) separation concentric HIDiC plant, in which consisted of about 29 m in height and 254 mm in diameter. Predicted simulation results such as concentration, temperature and flow rate of top and bottom in the rectifying section and stripping section show good agreement with observed data. The effect of mass transfer and heat transfer in the HIDiC is also discussed. Moreover, a simulator using this model is developed to multicomponent HIDiC with concentric and compact heat exchanger systems.

Introduction

Energy-saving process is the most promising next-generation technology for distillation with acceptable environmental characteristics and inexpensive process cost (Humphrey, 1995). So far, the research of separation process was usually carried out by a conventional distillation process. In the conventional distillation process is resources and energy-intensive industries because of adding a large amount of thermal energy as a separation agent. For this reason, several energy-saving methods such as intermediate heat exchanger process (Agrawal and Herron, 1998), multiple effect distillation process (Tyreus and Luyben, 1975) and prefractionation arrangement process (Ognisty, 1995) have been developed to overcome the disadvantage of conventional process. However, these involve the complicated process to obtain the energy saving and need initial cost. As a result, that complicated processing restricts the industrialization for energy saving technologies.

Recently, it has been reported that the energy saving with respect to elementary technology and zero reflux and/or no reboil operation can be obtained by using a Heat Integrated Distillation Column (HIDiC) (Nakaiwa *et al.*, 2003, Iwakabe *et al.*, 2004, in press.). The HIDiC has a similar structure to a double pipe heat exchanger. The rectifying section is contacted with the stripping section through walls in order to allow heat exchange between these two sections. Thus the residual heats in the column can be utilized in itself.

In this study, we have developed the HIDiC models by the high performance process simulator "gPROMS". The as-developed models have been used as bench plant for HIDiC, and their separation performance and energy-saving ratio have also investigated.

Simulation model

A physical model of a typical non-equilibrium stage model used Krishnamurthy and Taylor's method by the Maxwell-Stefan approach (Krishnamurthy and Taylor, 1985a, 1985b, 1985c). The following simplifying assumptions are made for the present simulation.

1) The condenser and the reboiler are assumed to be at equilibrium.
2) Vapor-liquid equilibrium occurs at the vapor-liquid interface.
3) Interfacial area for mass and heat transfer is equal to apparent surface area of the packing.
4) Bulk phase is perfect mixing.

The liquid hold up for structured packing is calculated using the empirical correlation developed by Xu *et al.*, (1997)

$$h_d = 0.0336 u_G^{1/3} u_L^{0.429} \tag{1}$$

The binary mass transfer coefficients are estimated using the empirical correlation developed by Bravo *et al.*, (1982a, 1985b):

$$\frac{k_G d_e}{D_G} = 0.0338 \left\{ \frac{d_e \rho_G (u_{Ge} + u_{Le})}{\mu_G} \right\}^{0.8} \left(\frac{\mu_G}{\rho_G D_G} \right)^{1/3} \tag{2}$$

$$k_L = 2\sqrt{D_L u_{Le} / (\pi S)} \tag{3}$$

Heat transfer coefficients are computed by using the Chilton-Colburn analogy (King, 1980). The resulting differential and algebra equations in this model are implemented in gPROMS.

Outline of bench plant

In order to show the validity of the present simulation model, comparison of the results by the present model with experimental data by a bench-scale HIDiC was made. **Figure 1** show a schematic diagram of the test plant. The test plant was a double pipe heat exchanger system, which consists of shell and tube column. This plant was 16m in height and 254mm in diameter. The shell side was a stripping section, which was operated at nearly atmospheric pressure with Mitsubishi Chemical MC Pack500. On the other hand, the tube side was a rectifying section. The rectifying section was operated at about pressure of stripping section times two. The diameter of tube side section changed to three stages from the top to the bottom. Their diameters were 140, 165 and 190mm and their lengths were 3, 8, 5m with MC Pack350, respectively.

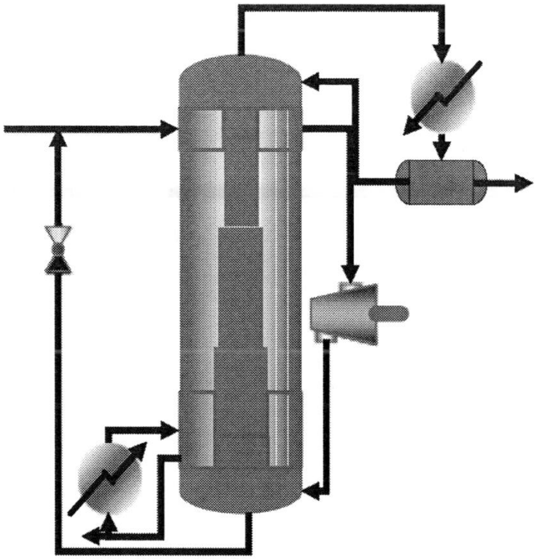

Fig.1 Schematic diagram of the HIDiC

In our previous paper, experimental approaches to zero reflux and/or no reboil operation in HIDiC of the benzene-toluene system (Naito *et al.*, 2000.) were made. This work show that the bench type HIDiC was more 40% of energy saving than conventional distillation column. The purpose of the present work is to develop a method for prediction of separation performance of HIDiC with structured packing by applying rate-based model.

Table 1 Operating condition for the tube unit of bench-plant

Pressure of the rectifying section	P_r	0.24	[MPa]
Pressure of the stripping section	P_s	0.124	[MPa]
Feed flow rate	F	340	[kl/h]
Reflux ratio	R	0	[-]
Feed composition	x_B	0.5851	[-]
	x_T	0.4148	[-]
Feed thermal condition	q	1	[-]

Comparison with experimental data

Table 1 summarized operating condition for the bench plant which was used for comparison purpose. **Table 2** shows a comparison of the predicted data, such as temperatures, flow rates and concentrations of top and bottom, by present model with observed ones. Good agreement is observed between the data and the present model, such as temperatures, flow rates and concentrations, respectively. In this results, present model with rate-based can be precisely predict separation performance of HIDiC.

Typical phenomenon of HIDiC

Figure 2 shows the predicted flow rate profiles along the column. The ordinate is the mass flow rates of vapor and liquid, and the abscissa is the packed height. It can be clearly seen that, the vapor and liquid flow rates increases as from the bottom to the top in the stripping section. In

Table 2 Comparison of simulation results with observed data in bench-plant

	Temperature				Flow rate		Concentration			
	The rectifying section		The stripping section		Bottom product	Distillate product	The rectifying section bottom		Bottom	Top
	Bottom Liquid	Top Vapor	Bottom Liquid	Top Vapor			Benzene	Toluene	Toluene	Benzene
	$T_{r,b}$	$T_{r,t}$	$T_{s,b}$	$T_{s,t}$	W	D	$x_{r,b,B}$	$x_{r,b,T}$	$x_{W,T}$	$x_{D,B}$
	[K]				[kl/h]		[-]			
Experimental	390.3	381.7	388.8	365.1	160	180	0.6226	0.3773	0.9984	0.9932
Calculated	395.9	388.5	385.5	370.6	152	188	0.6326	0.3674	0.9637	0.8835
Deviation [%]	1.43	1.78	-0.85	1.51	-5.12	4.56	1.61	-2.62	-3.48	-11.05

rectifying section, on the other hands, the flow rates decreases as from the bottom to the top. It is exhibited the evaporation of liquid in the rectifying section and the condensation of vapor in the stripping section. Their phenomenon was occurred because of heat exchange along the tube unit between rectifying and stripping sections. Note that the liquid flow rate at the top of the rectifying section is zero, i.e. there is no reflux.

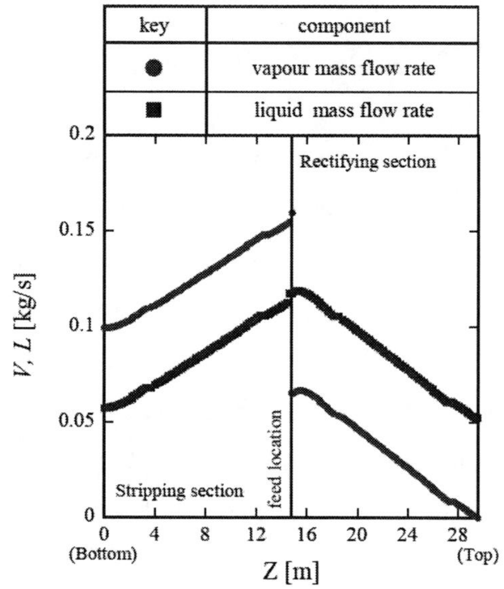

Fig.2 Variation of mass flow rates along the tube unit

Reduction of energy consumption

The start-up dynamic simulation quantified the pressure and heat exchange during the transition from cold and empty to full operation. **Fig. 3** shows the results. This verified that the equipment could be started up as planned. Further runs could then be performed to minimize the startup time to full production. The plots above show the energy input into the system through start-up to steady state operation. It can be clearly seen that the reboiler and condenser duties for both HIDiC configurations are substantially less than for the conventional column, while the additional compression energy requirement is relatively small. These indicate that the HIDiC system can save to 40% of the energy required for separation.

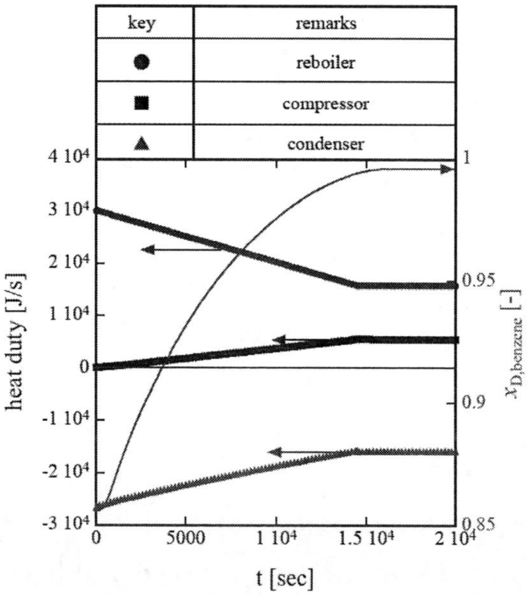

Fig.3 Transient responses of the startup operation.

Conclusions

The HIDiC with structured packing models were successfully simulated by using a rate-based model by Maxwell-Stefan equation for bench-scale plant. Simulation of separation performance of a bench plant by HIDiC with structured packing are carried out and the following results are obtained.

1) A method for prediction of separation performance of the HIDiC is developed used to rate-based model.
2) Comparison of predicted separation performance of the bench-scale HIDiC by present model shows an excellent agreement with experimental data.
3) For the separation of the binary mixture, the bench HIDiC showed about 40 % reduction of energy consumption than the conventional distillation column.

Nomenclature

D	binary diffusion coefficient, m^2/s
d_e	equivalent diameter of channel, m
F	feed flow rate, kl/h
h_d	dynamic liquid holdup, m^3/m^3
k	binary mass transfer coefficient, $kmol/m^2s$
L	liquid phase mass flow rate, kg/s
P	pressure, MPa
q	thermal condition
R	reflux ratio
S	channel side dimension, m
T	temperature, K
u_G	superficial gas velocity, m/s
u_L	superficial liquid velocity, m/s
V	vapor phase mass flow rate, kg/s
x	liquid phase mass fraction
y	vapor phase mass fraction
Z	packed height from top of column

Greek symbols

μ	viscosity, Pa·s
ρ	density, kg/m^3

Subscripts

B	benzene
b	bottom
G, V	vapor phase
L	liquid phase
r	rectifying section
s	stripping section
T	toluene
t	top

Acknowledgements

This work was supported by New Energy and Industrial Technology Development Organization (NEDO).

Literature Cited

Agrawal, A. and Herron, M., 1998, Intermidiate reboiler and condenser arrangements for dibary distillation columns , *AIChE J*, 44: 1316-1324.

Bravo, J. L. and Fair, J. R., 1982, Generalized correlation for mass transfer in packed distillation columns, *Ind Eng Chem Process Des Dev*, 21: 162-170.

Bravo, J. L., Rocha, J. A. and Fair, J. R., 1985, Mass transfer in gauze packings, *Hydrocarbon Processing*, 64: 91-95.

Iwakabe, K., Nakaiwa, M. Huang, K. Nakanishi, T. Ohmori, T. Endo, A. and Yamamoto, T., 2004, Recent advances in the internally heat-integrated distillation columns (HIDiC), *AIChE Annual meeting 2004*, 227f : *In Honor of Professor Johann Stichlmair-*, Austin, Texas, USA.

Iwakabe, K., Nakaiwa, M. Huang, K. Nakanishi, T. Røsjorde, A. Ohmori, T. Endo, A. and Yamamoto, T., Energy savings in multicomponent separation using an internally Heat-Integrated Distillation Column (HIDiC), *Applied Thermal Engineering*, accepted.

Humphrey, J. L., 1995, Separation processes: playing a critical role, *Chem Eng Prog*, 91(Oct): 31-41

King, C. J., 1980, Separation Process (MacGraw-Hill, New York, USA).

Krishnamurthy, R. and Taylor, R., 1985, A Nonequilibrium stage model of multicomponent separation processes, part I: Model description ad method of solution, *AIChE J*, 31: 449-456.

Krishnamurthy, R. and Taylor, R., 1985, A Nonequilibrium stage model of multicomponent separation processes, part II: Comparison with experiment, *AIChE J*, 31: 456-465.

Krishnamurthy, R. and Taylor, R., 1985, A Nonequilibrium stage model of multicomponent separation processes, part III: The influence of unequal component-efficiencies in process design problems, *AIChE J*, 31: 1973-1985.

Naito, K. Nakaiwa, M. Huang, K. Endo, A. Aso, K. Nakanishi, T. Noda, H. and Takamatsu, T., 2000, Operation of a bench-scale ideal heat integrated distillation column (HIDiC): an experimental study, *Compt Chem Eng*, 24: 495-499.

Nakaiwa, M., Huang, K. Endo, A. Ohmori, T. Akiya, T. and Takamatsu, T., 2003, Internally heat-integrated distillation columns: A Review, *Chem Eng Res Des*, 81: 162-177.

Ognisty, T. P., 1995, Analyze distillation column with thermodynamics, *Chem Eng Prog*, 91(Feb): 40-46

Tyreus, B. D. and Luyben, W. L. 1975, two towers cheapter than one?, Hydrocarbon Processing, 93: 93-96

Xu, X. Zhao, Z. and Tian, S., 1997, Study on catalytic distillation processes. Part III. Prediction of pressure drop and holdup in catalyst bed, *Chem Eng Res Des*, 75: 625-629.

58a On Trays, Packings, and a Life in Distillation

Jose Bravo, Shell Global Solutions, 3333 Hwy 6 S., Houston, TX 77082

Jose will deliver an excellent presentation that will stimulate the intellectual palate of all those in attendence in honor of our good friend John Farone.

58b Different Strokes: a Review of the Various Unique Devices Tested at Fractionation Research, Inc

Mark W. Pilling, Sulzer Chemtech USA, Inc., 4019 S. Jackson Ave., Tulsa, OK 74107

Over the years, a variety of interesting devices have been tested at Fractionation Research, Inc. Some of these devices have performed extremely well and others have been less successful. This paper will review the various devices, focusing on their unique characteristics and discuss the likely explanations for their performance.

58c Evaluation of Structured Packing Hydraulic Performance Using X-Ray Computed Tomography

Christian W. Green, The University of Texas at Austin Separations Research Program, 104A Franklin Blvd., Austin, TX 78751 and R. Bruce Eldridge, Chemical Engineering, University of Texas at Austin, 1 University Station C0400, CPE 4.408, Austin, TX 78712.

Existing models used to predict hydraulic and mass-transfer performance of packed columns are semi-empirical and rely on column-average properties for parameters such as liquid holdup. These models do not accurately describe the performance observed in operating packed columns, particularly for hydraulic studies. X-ray Computed Tomography (CT) is a non-invasive, non-intrusive imaging technique that was used to provide insight into the liquid distribution and other flow properties within an operating air-water contactor utilizing structured packing. Several commercial stainless steel structured packings, including Sulzer Mellapak 250Y, were installed in a 6-inch air-water column and imaged under irrigated and static conditions using X-ray CT. Liquid rates varied from 0-20 GPM/ft2, and vapor rates ranged from F-factor 0 to 2. Programs to analyze the resulting CT images were written using IDL, a commercial image processing code. Liquid holdup was calculated at several elevations within the packed section through analysis of CT slices. Holdup measurements from slices obtained at evenly spaced intervals were integrated over the height of the column and compared to both experimental data and existing hydraulic performance models. Hydraulic behavior at packing element joints was further investigated by taking several CT slices at close intervals around the joints. Liquid holdup profiles of the packed section showed local peaks at or slightly above packing element joints. Three-dimensional models up to 6" in height of both dry and irrigated packing were obtained by taking a series of CT slices at close intervals and reconstructing using 3-D imaging software. Use of 3-D modeling aims to provide detailed images of liquid location within a volume of the packing, yielding insight to liquid flow path, film thickness and behavior, and wetted and interfacial areas. Data from the CT experiments is to be incorporated in the development of a new predictive model for packed column hydraulic performance.

58d Pressure Drop of Packed Column Internals

Armin Rix, Engineering Projektierung 1, Degussa AG, Paul Bauman Str. 1, Marl, 45764, Germany and Zarko Olujic, Delft University of Technology, Leeghwaterstraat 44, Delft, 2628 CA, Netherlands.
Because of their low pressure drop per theoretical stage random and particularly structured packings are frequently used in vacuum applications. For applications requiring operation at absolute pressures bellow 100 mbar usually total allowable pressure drop is specified. Demanding separations require frequent collection and redistribution of liquid, and this means dealing with a source of pressure drop which cannot be ignored as it is usually done, taking into account the pressure drop of the irrigated packed bed only. Unfortunately only scarce data on the pressure drop of liquid and vapor redistribution sections can be found in open literature. The purpose of this paper is to summarize available pressure drop data for most common types of liquid collectors and distributors and to evaluate and discuss them in a systematic way in terms of overall pressure loss coefficient which could provide a practical guidance to people dealing with design and rating of packed columns. Main body of experimental data evaluated in this study has been obtained using 1.4 m internal diameter column hydraulics' simulator available at the Delft University of Technology.

A Deep-Knowledge Based EVOP Approach for Distillation Process Optimization

Lowell Pless* and Simon Xu
Process Diagnostics, Quest TruTec, 11005 West Fairmont Pkwy, La Porte, TX 77571

* Contact: Email l.pless@QuestTruTec.com, Phone: (281) 471-8970, Fax: (281) 471-0078

Presented in Topical 8: Distillation Topical - #56 - Distillation Honors: John Farone of the 2006 Spring A.I.Ch.E. Meeting, Orlando, FL, April 23-27, 2006

UNPUBLISHED

A Deep-Knowledge Based EVOP Approach for Distillation Process Optimization

Lowell Pless* and Simon Xu,
Process Diagnostics, Quest TruTec,
11005 West Fairmont Pkwy, La Porte, TX 77571

* Contact: Email l.pless@QuestTruTec.com, Phone: (281) 471-8970, Fax: (281) 471-0078

Introduction

Evolutionary Operation (EVOP) is a methodical experimental approach to optimize or improve continuous processes [1,2]. EVOP uses designed experiments that vary process variables within existing operational constraints to improve the process continuously. We are most familiar hearing the term Six Sigma with the application of designed experiments with regard to discrete manufacturing processes such as the computer chip industry. But the disciplined thought process has also been applied to process industries such as our refining and petrochemical plants.

This paper will present a deep-knowledge based EVOP approach for separation process optimization, which involves uncertainties in predicting gas-liquid contacting hydraulics and mass transfer. A case study has been used to illustrate the usefulness of this experimental method supplemented with gamma scanning and tracer techniques in solving a complex process operating problem involving an absorption process.

The basic procedure for the Evolutionary Approach is:

- Choose a base operating point
- Evaluate the objective function at base point and a set of regularly spaced operating points around the base point
  For 2 variables, form a square around the base point
  For 3 variables, form a cube
- Move the process to the operating point(s) that has the most potential for maximizing the objective function
- Use this point as the new base point
- Repeat steps 2 through 4 until no further improvement occurs

Troubleshooting a Revamped Absorber

A petrochemical company had revamped a packed column in order to improve its separation efficiency.

The project goal was to increase the operating temperature of the lean water to save energy costs by a reduction of refrigeration. At the same time the overhead product purity had to be maintained to achieve environmental compliance and minimize product loss.

It is well known that the absorption column will need more theoretical stages for the same separation at a higher operating temperature. Replacement of existing packed beds with high efficiency packing was believed to be able to deliver sufficient stages. The revamp consisted of replacing the top and middle beds of packing with higher efficiency packing and replacing the liquid redistributor.

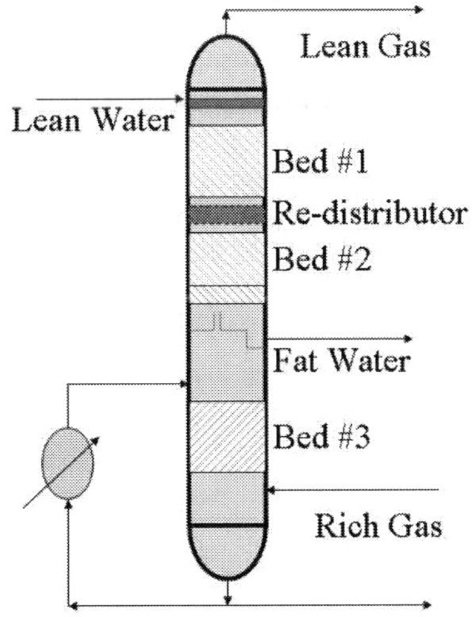

Fig. 1 – simplified process diagram of absorption system

An improvement in mass transfer efficiency, in theory, would allow the absorber to operate at a higher temperature thereby using less or no refrigeration of the lean water stream and saving energy.

Problems ensued immediately after startup. The Absorber column was operating with a higher than expected ΔP and the overhead product was off specification. A quick troubleshooting gamma scan was done to diagnose the problem.

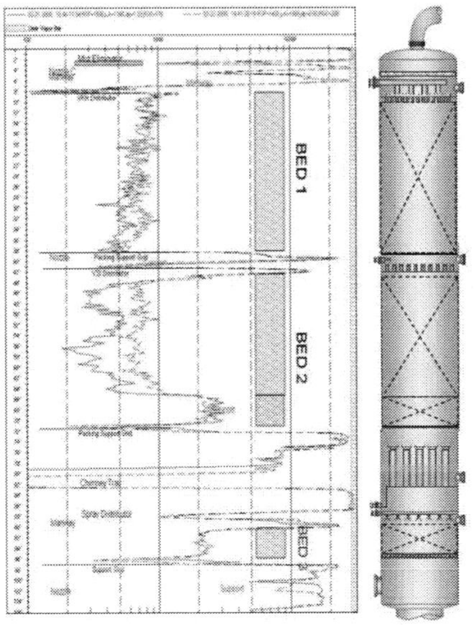

Fig. 2 – Results of scan through middle bed.

The scan (results in Fig. 2) showed wildly varying process densities down through the middle bed of packing. This was reminiscent of foaming as you have a regime of light density where the process liquid is highly aerated or foamy with some excess liquid building in the packing above (showing as an area of higher density on top of lighter density material) [3]. This particular process is a known foaming system and contaminants would exacerbate the problem. The decision was made to circulate cleaning solution through the Absorber to wash any contaminant or other foaming agents out of the system. Over the course of a week subsequent scans and chemical analyses were done to determine when the system was "clean".

The Absorber column was restarted and while the ΔP was more near its normal range the overhead product was still off specification. The next issue that was raised was to question the liquid distribution - one of the most critical aspects with regard to a packed column operating satisfactorily. To answer this question a grid scan was performed.

To evaluate the liquid distribution in packed columns a special application of gamma scanning developed specifically for packed column is used [4]. This scanning procedure uses four scan lines, in a 2x2 grid with equal cord lengths, are used as shown in ***Figure 3***.

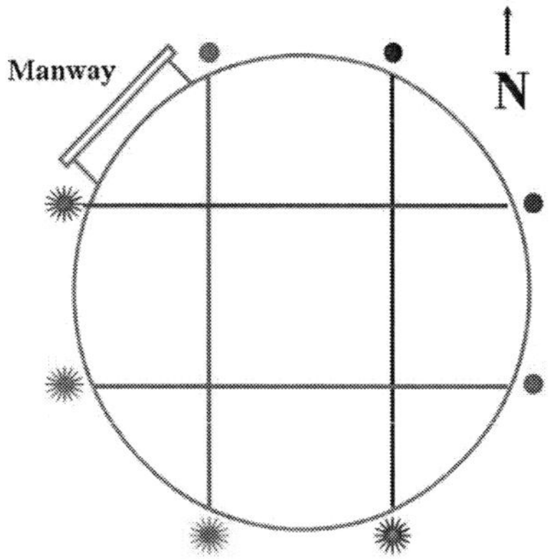

Fig. 3 – Orientation for Tru-Grid™ Scan

It is critically important to have equal scan chord lengths so that the radiation passes through an equal thickness of "material" on all sides of the column. The basis is that uniform or good liquid distribution also has uniform or equal bulk density through the packing. Therefore when the data plots of all four scan lines coincide, or overlay on top of each other; this indicates uniform liquid distribution through the packing. Problems are diagnosed when non-uniformity is observed from the scan data plots. Any divergence among the scan data can be an indication of maldistribution. Lower "radiation intensity" indicates excess liquid and higher "radiation intensity" is associated with liquid deficiency.

The scan results showed some interesting results.

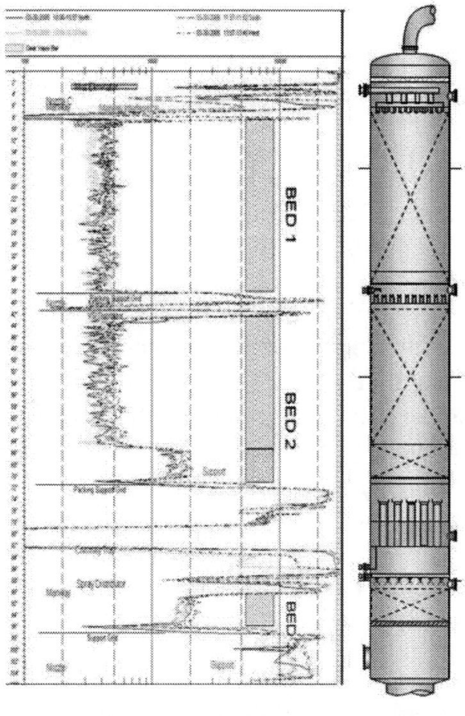

Fig. 4 – Results of grid scan.

The liquid through the middle bed was mal-distributed. There were signs of some foaming through the packing so this could be contributing to the mal-distribution. But even at the top of the bed the liquid never was distributed well from the liquid redistributor. However before blaming the liquid redistributor, which was the popular choice, one needs to explain further what is happening at the top of the bed.

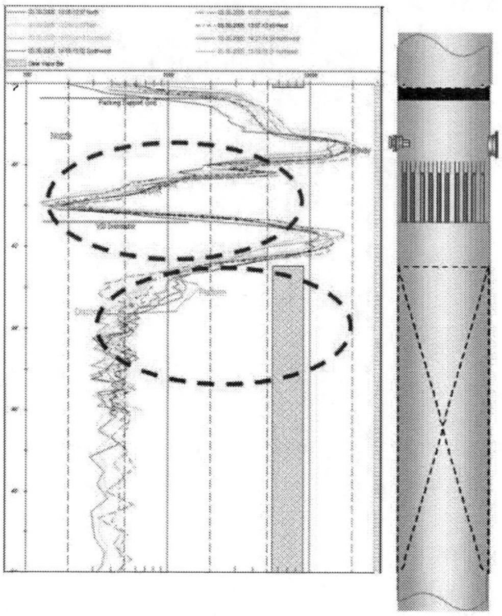

Fig. 5 – Close-up of top of middle bed and liquid redistributors.

The scan profile of the top layer of the packing did not appear "normal". One can see a large difference in radiation intensity among the four scan lines. (Fig. 5) The question posed was, "Is this density difference in the top of the packing due to liquid maldistribution or is there damage to the packing?" The grid scan is a useful screening tool, but not a precise measurement of process densities. However we do have a tool to use to map out the process densities of the internal process in a packed tower, known as CAT-scan [5].

The sequence of applying a grid-scan and a CAT-scan is similar to the process that a medical doctor might use. The grid scan is a quick, relatively inexpensive, reliable screening tool much like a medical X-ray to find gross problems within a packed column. When more detailed information is needed to solve a problem we use a Cat-scan much like the medical doctor would follow an X-ray with an MRI or a CAT-scan.

Gamma-ray tomography takes advantage of the penetrating property of high-energy gamma radiation to "detect" internal process densities of reactors, absorbers and distillation columns. In our experience, data collected from 20-200 scan lines are generally sufficient to construct a representative tomography of most industrial packed towers.

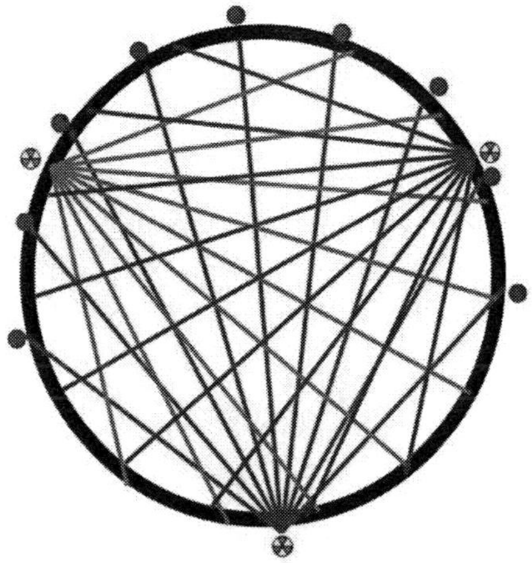

Fig. 6 – Simplified example of scan line orientations for a CAT-Scan.

Figure 6 shows a fan-beam scheme of 3 fan rotations, with 9 scan paths per each rotation (a 3x9 pattern). Most typically on industrial packed towers a 9x9 pattern is used. The objective is to optimize the placement of the scan paths to produce the best image. A proprietary program has been developed to calculate source selection, source/detector placement, data reconstruction, and 3D-image development. Data reconstruction is done with a 4th order polynomial density function containing 15 coefficients that must be determined by regression of the scan data. [5]

Before taking corrective action the decision was made to use CAT-scans to assemble more information about what was happening at the top of the middle bed.

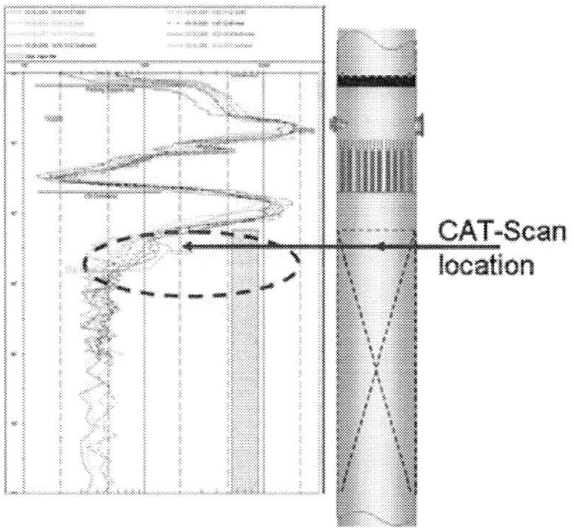

Figure 7 above shows one location where a CAT-scan was used.

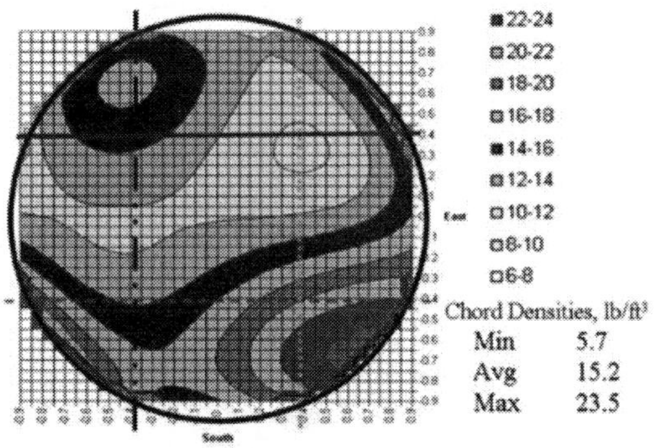

Fig. 8 – Results from CAT-Scan with chord densities

The results of the CAT-scan (***Fig. 8)*** shows that a portion of the top layer of packing had density equal to the dry bulk density (5.7 lb/ft^3) of the packing itself, while another section had a very high density. The chord densities calculated from the CAT-scan measurements ranged from a minimum of 5.7 lb/ft^3 to 23.5 lb/ft^3. The most logical explanation that could give a combined packing material plus process density equal to the dry packing density would be missing or damaged packing. This was a difficult conclusion for some to accept. Then some additional details were discovered. A close examination of the recorded operating data showed a pressure surge during the time the Absorber was operating with the foaming problem. The hardware vendor calculated that the pressure surge, given that there was no hold-down plate, was enough to theoretically dislodge the top layer of packing.

Still there were those that thought the liquid maldistribution problem was due to a malfunctioning liquid redistributor, so further attention was paid to the liquid redistributor. A series of scans were done across the redistributor to measure the liquid height on the redistributor. (Results in Figures 5 and 9)

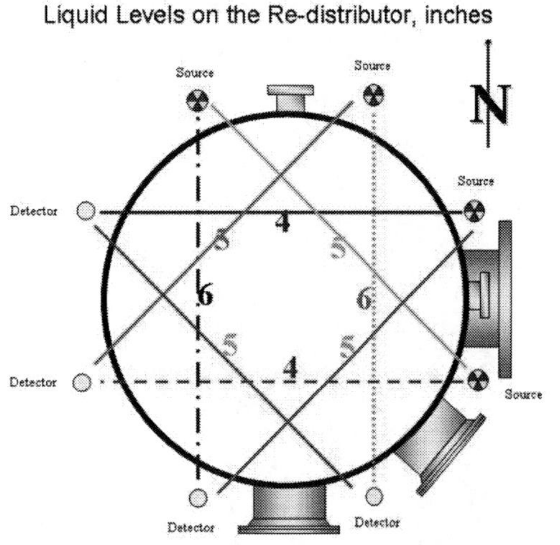

Fig 9. – Scanlines used and liquid levels measured on liquid Re-distributor.

The results of measuring the liquid level on the redistributor showed that the liquid level was uniform and close to the expected height as prescribed by the hardware vendor. There was no new compelling evidence, especially in light of the CAT-scan results, that would indicate a problem with the liquid redistributor. We remained convinced the immediate problem was the indicated damage to the top layers of packing. Finally enough people decided that the Absorber needed to be shut down, inspected, and repaired.

Figure 10 (above) shows a picture of the liquid redistributor immediately after opening the manway. The liquid redistributor was secure, level, and chimney hats were in place.

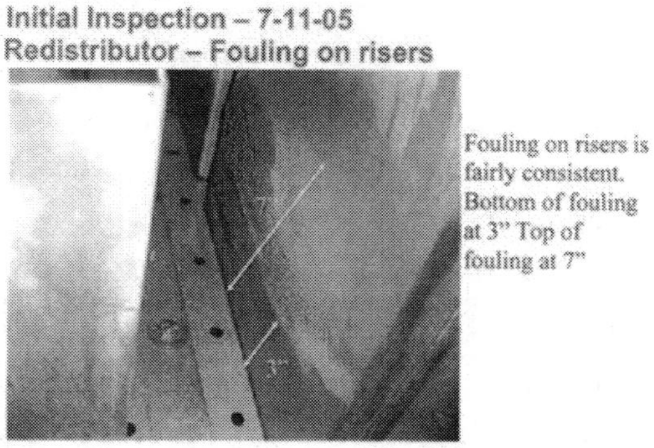

Figure 11 (above) shows the water marks left from the water absorbent and shows that the water level varied from 3 to 7 inches which matches the measured 4 – 6 inch water levels measured by gamma scanning. ***Figures 12 through 15*** (shown below) are pictures of the damage to the top layers of the middle bed packing.

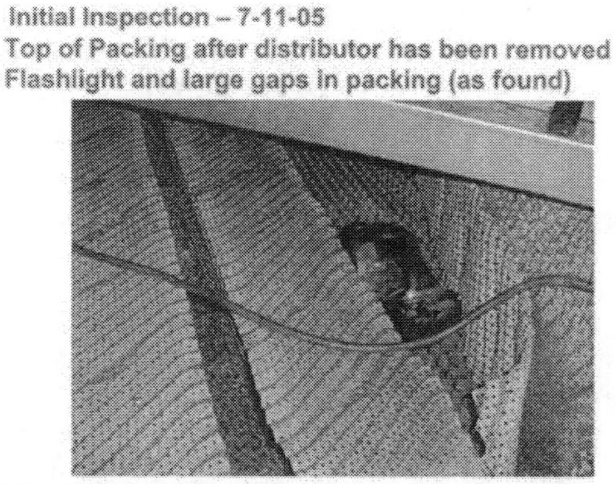

Fig. 12

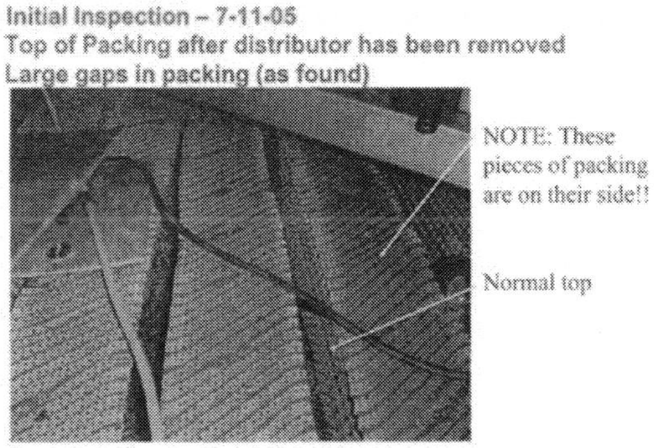

Fig. 13

Fig. 14

Fig. 15

The Absorber was repaired and restarted. While the absorption process was improved the overhead product was still off specification – averaging 350 ppm of the key component versus a target of less than 100 ppm. So, what was next? The process simulation for the Absorber column showed that on specification operation would need 12 theoretical stages from Beds 1 and 2. The purpose for Bed 3 was primarily for heat transfer and secondarily for washing. Current performance of the Absorber indicated that only 6 – 7 theoretical stages were being obtained.

EVOP for Optimizing the Process

Up to this point the EVOP approach had not really been used because major problems with the physical (mechanical) condition of the Absorber had been identified and needed correction first. Now that the Absorber was in good physical condition but still not meeting expectations the EVOP technique was utilized.

Following the recommended steps for the EVOP evaluation taken from references [1, 2]:

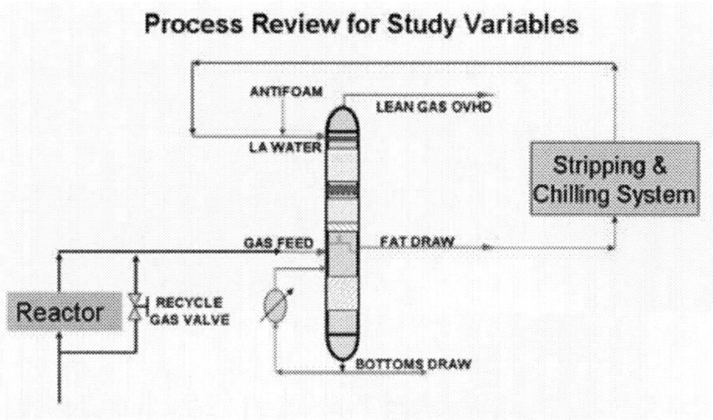

Fig. 16

1. Choose 2 or 3 variables of interest that can be changed and can be tightly controlled.

   The variables chosen for the experiments must have some potential for improving the process. ***Refer to Figure 16 shown above.*** Temperature and L/V ratio have a great impact on the absorption process with lower temperatures and higher liquid flow tending to favor the process efficiency. In addition foaming was a big concern in this case so antifoam addition was a variable added to the study. Lastly gas flow was added as another option to optimize the L/V ratio.

2. Set "high" and "low" conditions for each of the variables.

   Operations were consulted and the following limits on the experimental variables were agreed to:

   - LA (lean absorbent) water rate:± 5%
   - Gas Feed Rate: ± 5%
   - Anti-foam Injection: 1 – 5 gpd
   - LA Water Temperature: ± 3%

   Based on the process simulation these ranges would give unambiguous results on the overhead product.

3. Operate at each combination of conditions.

   The EVOP technique calls for manipulating 2 or 3 variables at one time. Since four variables had been identified to be of interest the gas flow rate was temporarily set aside to concentrate on the other three first. Also the gas flow rate was not as "independent" as the other three since the gas feed rate to the Absorber was set by the reactor rates and had less "degree of freedom". So the following table shows the design of the first experiment which would start with the antifoam test. If the Absorber had a foaming issue, which gamma scans seemed to suggest, then this would need to be optimized first.

Process Variable	Process Units	Low Value	Current Value	High Value
LA Water Flow	Million lb/hr	1.7	1.8	2
LA Water Temperature	°F	70	72	74
Anti-foam Rate	GPD	1	2	5

Run 1: Antifoam Test

Gas feed rate and LA temperature were kept constant while LA water flow and anti-foam rates were varied over the four sets of conditions (high – high, low – low, high – low, low – high, respectively). The results showed that whether at high or low anti-foam addition rates the key component in the overhead remained in the 350 ppm range. Therefore either the anti-foam chemical did not work or the system was not foaming as had been expected. The conclusion was to cut the anti-foam to a minimum rate and save the chemical cost.

Run 2: LA Temperature vs. Gas Flow Rate

The variables of LA water temperature, Absorber gas feed rate, and overhead composition interact with each other because of heat integration within the process.

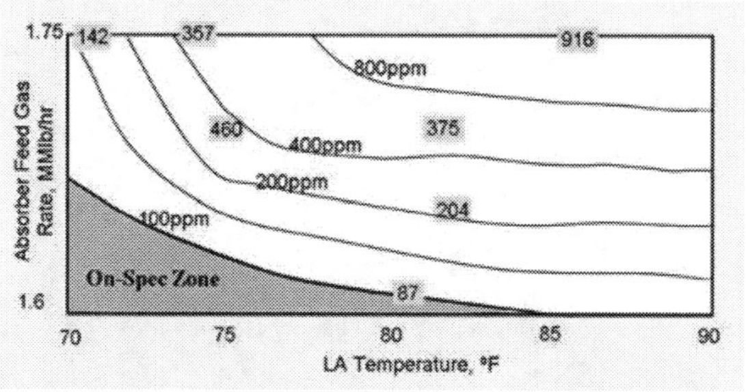

Fig. 17 Overhead key component concentration vs. LA Temperature and Absorber Gas Feed Rate

Approximate contours (***Figure 17***) for these variables have been drawn by interpolation from process simulation results with experimental points shown confirming the general shape of the contour. The contours were helpful to identify the optimum combination of the variables for the lowest overhead impurities. The process was able to reach the overhead specification of the key component being less than 100 ppm at proper variable settings. However it appeared that to maintain this specification the process throughput had to be several percent less than the design capacity. Further investigation indicated that the gas volumetric flow to the absorber depended significantly on the feed gas composition, which is a mix of reactor effluent gas and recycle gas. Operations were able to reduce the volumetric flow rate by adjusting the component concentrations in the recycle gas allowing the process throughput to reach design capacity with the overhead specification met.

Run 3: Mass Balance and Leak Detection

During the EVOP tests, Operations reviewed the mass balance of the process and raised the question about the amount of gas flow into the reactor. It was estimated that some gas feed might be leaking or "shortcutting" through the gas by-pass line, which was used for by-passing the reactor during startups (see *Fig. 16 on page 7*). A radioactive tracer test was performed and confirmed that about 15% of feed gas was by-passing the reactor because of the valve malfunction.

Runs 4 and More: Continuous Optimization

With more systematic process data collected during the EVOP tests, it was found that the whole process, including the reactor and the absorber, could be further optimized by more strategic design of experiments. The variables shall include extensive parameters, like the catalyst life, reaction conversion, recycle rate, and weather temperature.

Closing Remarks

Gamma Scans can provide helpful insights on mechanical integrity and hydraulics of gas-liquid contacting columns. In this particular case the deep-knowledge on column internal operations enhanced the EVOP approach for separation process optimization, which involved uncertainties in predicting gas-liquid contacting hydraulics and mass transfer. A case study has shown the usefulness of this experimental method supplemented with gamma scanning and tracer techniques in solving a complex process operating problem involving an absorption process.

References

[1] Box, G. and N. Draper, Evolutionary Operation – A Statistical Method for Process Improvement, John Wiley and Sons, New York, 1998.

[2] Moffatt, S. and V. Monical, Using Evolutionary Operation (EVOP) to Optimize Distillation Operations, A.I.Ch.E 2005 Spring Meeting, Atlanta, GA, 2005.

[3] Xu, S. and L. Pless, Understand More Fundamentals of Distillation Column Operation from Gamma Scans, A.I.Ch.E 2001 Spring Meeting, Houston, 2001.

[4] Asseln, B. and L. Pless, "Using Gamma Scans to Plan Maintenance of Columns," Petroleum Technology Quarterly, Spring 2002, p.115.

[5] Mixon, W. and S. Xu, Identify Liquid Mal-distribution in Packed Distillation Towers by CAT-Scan Technology, A.I.Ch.E 2005 Spring Meeting, Atlanta, 2005.

Extractive Distillation
A Process Improvement Project

James P Farone, Senior Process Engineer, Trimeric Corporation, Buda, TX

Menno Kroes, Process Technology Manager, Huntsman International, The Woodlands, TX

Mark Mueller, Senior Research Associate, Huntsman International, The Woodlands, TX

Abstract

This paper gives a high level review of the technical methodology used to improve the overall performance of an extractive distillation purification unit that was complicated by unusual chemical equilibria. The general process, project technical difficulties, technical methodology used to troubleshoot the system, solution implementation barriers and implementation results are described.

The extractive distillation unit in question produces a high volume overhead product at >99.98% purity. The main focus and challenge of the project was to improve the extractive removal of a critical impurity present at low concentrations (~50 ppm) without sacrificing the removal efficiency of other higher concentration impurities in the system. As an added challenge, in this particular extractive distillation system, process simulation accurately predicts the system performance of all impurities with the exception of the critical impurity.

Note: All chemical and process specific information is generalized to protect the intellectual property of all parties involved. This paper focuses on the methodology used to improve the extractive distillation process and not the details of the extractive distillation process itself. All three authors were Huntsman Corporation associates at the time of this project. Although Trimeric Corporation performs process trouble-shooting projects of this type, Trimeric was not involved in this specific project.

Process Overview

An existing Huntsman Corporation production unit produces a high volume product to 99.98 wt% purity. This production unit is divided into four major processes; feed preparation, reaction, purification, and process utilities. Product purification is achieved using a series of distillation columns for removal of impurities while minimizing product loses. One key purification step involves an extractive distillation process consisting of a solvent contact column and a solvent regenerator.

The solvent contact column and the solvent regenerator are classic distillation columns which include a reboiler and an overhead condenser/reflux system. Raw product and lean solvent are fed to the contact column. Clean product leaves the contact column as an overhead product and rich solvent leaves as the bottoms stream. Rich solvent from the bottom of the contact column is fed to the solvent regenerator. Any product and impurities contained in the rich solvent are stripped from the solvent and leave the regenerator as an overhead stream. This overhead stream from the regenerator is used as a liquid fuel to produce energy in other areas of the plant. Lean solvent from the bottom of the regenerator is recycled back to the contact column.

Project Description

Objective

1. Reduce product loses in the Extractive Distillation Process.
2. Identify a method to improve the removal of the critical impurity without sacrificing the removal of other impurities from the product.

Background

The extractive distillation process involves four major chemical components; a desired product, a solvent, a critical impurity, and other impurities. The other impurities represent all chemical compounds in the contact column feed that are typically kept in control by the extractive distillation process. These impurities were identified during the original design of the extractive distillation process. The critical impurity represents an unexpected impurity in the process that must be controlled at a low concentration to maintain a high quality product. The extractive distillation process is the only identified exit point in the unit for the critical impurity other than the product stream.

Since the critical impurity was not anticipated in the original process design, no research had been done to identify the response of the critical impurity to operating conditions in this extractive distillation process. On a pure basis, the critical impurity would be considered a light component when compared to the other components in the system thus the first attempt to control the critical impurity was to reduce the contact column bottoms temperature. This had two effects. First, as would be expected, it resulted in additional product loss out the bottom of the contact column. Any product in the solvent leaving the contact column is lost to liquid fuel via the solvent regenerator.

Secondly, reducing the contactor column bottoms temperature had a marginal effect on the removal of the critical impurity. Since there was no operating history on this specific application of extractive distillation and the product purity appeared to be roughly in control, the process was operated in this mode until the issue could be investigated in greater detail. Initial plant observations indicated the removal of the critical impurity improved when the contact column bottom temperature was reduced and got worse when the temperature was increased. Figure 1 provides a process flow diagram with typical impurity concentrations during this mode of operation.

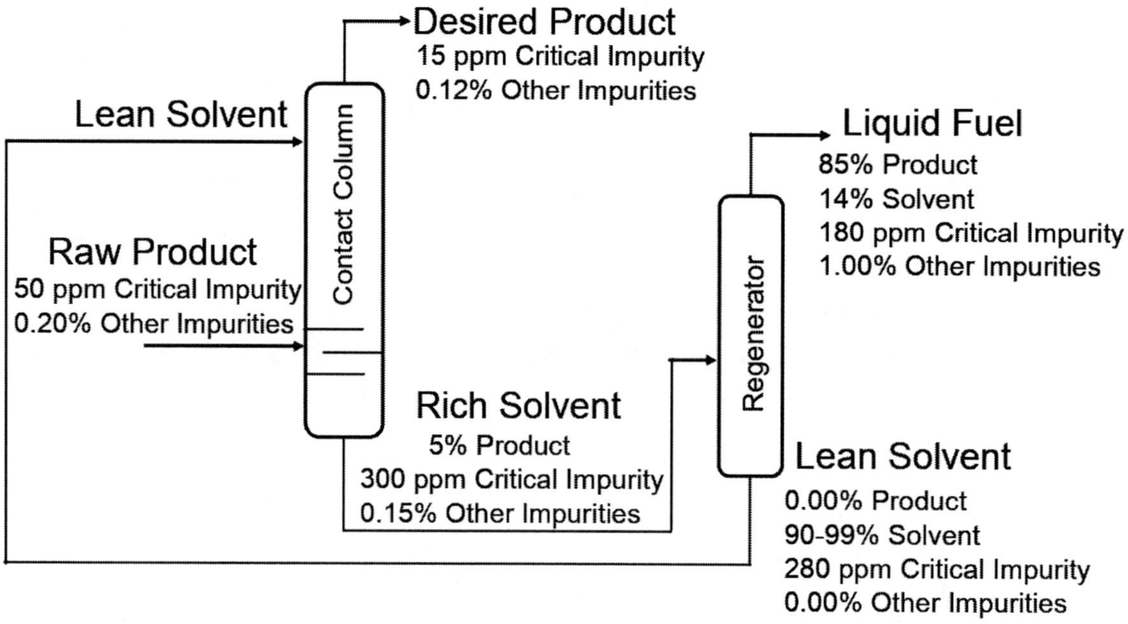

Figure 1: Extractive Distillation Process Flow Diagram

Based on these initial observations, operating practice was to reduce the bottom temperature during critical impurity excursions.

Problem Solving Methods Used

Initial Technical Approach

Since the relationship between operating conditions and the concentration profile of the critical impurity was unclear in this distillation environment, the first initial troubleshooting test was to add the critical impurity to the original process design simulation to determine if the simulation matches the current performance of the process with respect to the critical impurity. Unfortunately, the simulation was not able to predict the behavior of the critical impurity.

Process Model Development

Based on a rough comparison of the process simulation model vs. the performance of the distillation process, it appeared the simulation did a good job predicting the process performance of all components in the system with the exception of the critical impurity. However, the simulation had been developed in the design phase of the process development and had not yet been validated against the actual operating performance of the full scale process.

A validated model of the process was required to theoretically evaluate any process condition changes that could influence the critical impurity's Concentration profile. Plant operating and performance data were collected at various operating conditions for comparison to the simulator's predicted performance. In addition, the VLE used in the simulator were reviewed and updated to reflect the best data currently available.

As a result, a validated process model was developed for the extractive distillation process that was accurate over the range of operating conditions. This validated model provided a means to predict the system performance in regard to all impurities except the critical impurity.

Investigation of the Critical Impurity

Unavailability of a pure sample of the critical impurity and its low concentration in the extractive distillation system made experimental development of VLE data prohibitive. As an alternative, two parallel methods were used to gain insight into the relationship between the critical impurity and the distillation system. One was experimental evaluation of samples taken from the full scale operating process and the other was an evaluation of plant data at various operating conditions.

Samples were collected from the contact column bottoms stream and distilled in a batch distillation. Since the bottom stream from the contact column contained a significant amount of product, the batch distillation would provide information regarding the way impurities interact with the solvent. An interesting and unexpected result, shown in Figure 2, indicated that the critical impurity would not leave the solvent until all of the product had been stripped out. This was contrary to estimations based on relative volatilities of the components. The critical impurity was expected to strip out of the solvent before the product.

Therefore, as the extraction bottom temperatures are lowered, more product is allowed to get to the Regenerator. The product would go to the top of this column, make up most of the reflux and subsequently prevent the critical impurity from getting out. The critical impurity would become concentrated in the solvent and recirculate back to the top of the contact column giving the critical impurity another opportunity to exit with the product. One possible conclusion is that the removal of the critical impurity by the solvent may be dramatically improved if the contact column was operated to minimize product loss in the solvent stream.

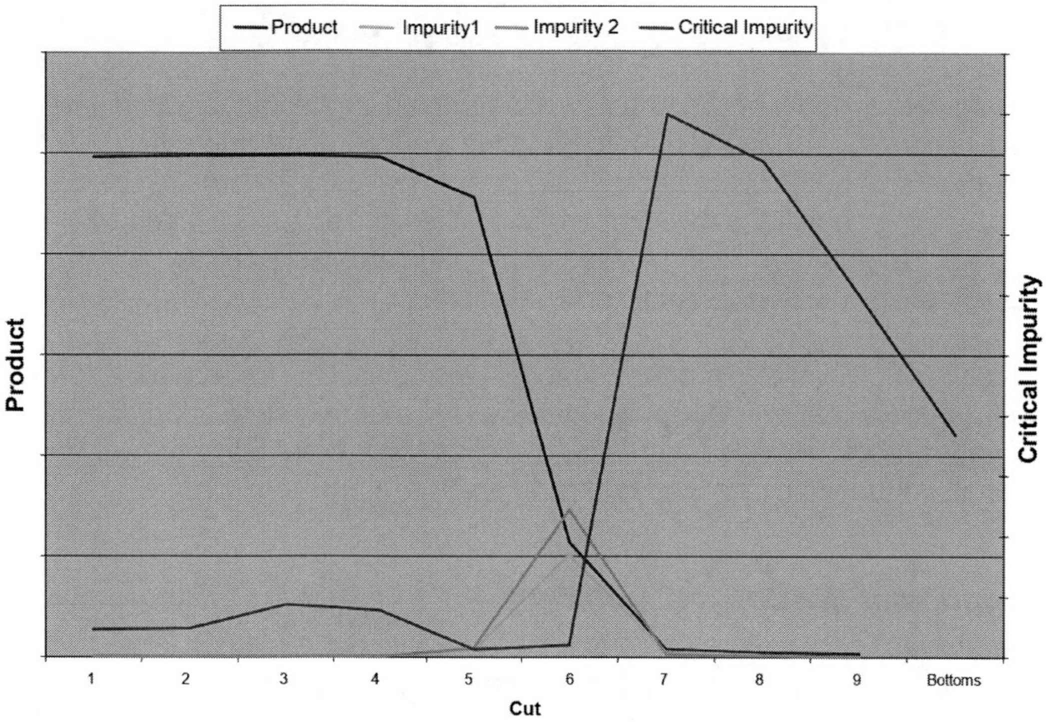

Figure 2: Overhead Cuts from Solvent Batch Distillation

In parallel to the batch distillation of the solvent, plant operating and performance data were collected and evaluated to identify any cause and effect relationships between the critical impurity and the operating conditions of the contact column. Evaluation of process data over a short time period (one hour) confirmed that the concentration of the critical impurity decreased when the contact column bottom temperature decreased. However, over a longer time period (multiple hours), the concentration of the critical impurity climbed back to the concentration prior to the temperature change. The exact opposite response is observed if the contact column temperature is raised, thus indicating the extractive distillation system experiences an accumulation/de-accumulation transition period after a temperature change is made on the process. This transition period lasts multiple hours before steady state is reached.

Based on the results from the batch distillation and plant data evaluation, it was postulated that the contact column bottoms could be operated at a higher temperature to avoid product loss thus preventing it from "capping" the regeneration column. Since it was shown that the product was stripped from the solvent preferentially to the critical impurity, it was also postulated that this mode of operation would not disrupt the steady state concentration of the critical impurity profile in the contact column. In fact, by avoiding the product "capping" in the regenerator, the critical impurity now has a way out of the process leading to a reduction of its concentration in the final product.

Validated Model Prediction at Low Product Loss Conditions

The next step was to use the validated process model to predict the approximate operating conditions needed to achieve minimal product concentrations in the solvent stream leaving the contact column. Once the needed operating conditions had been estimated, the process performance predicted by the model at these conditions was evaluated and found to be within acceptable tolerances.

Conclusions

- The existing process simulation model can reliably predict the system performance in regard to all impurities with the exception of the critical impurity.
- The critical impurity appears to have a greater affinity to the solvent than the product.
- Product concentration in the solvent stream to the regenerator must be minimized to avoid solvent "capping" thus allowing the regenerator to strip the critical impurity from the solvent.
- Predicted process performance at conditions required to minimize product losses is within acceptable product purity tolerances.
- Transition from current operating conditions to the new operating conditions must be made in a controlled stepwise manor to minimize concentration spikes of the critical impurity in the product.

Implementation Considerations

Since the contact column bottom temperature must be increased to reach the low product loss operating conditions and since the transition period during the temperature increase will lead to concentration spikes of the critical impurity in the product, good communication with operations and management was required prior to implementation. An implementation plan was developed requiring small temperature increases separated by a couple of days to allow the system to reach equilibrium between each step. In this way, the process reached the new operating conditions without any major upsets or operational issues.

Results

The greatest challenge for this project was the unusual relationship between the critical impurity and the extractive distillation process. In most circumstances it would be counter intuitive to raise the temperature and allow the impurity level to rise for a while in order to achieve a lower steady state impurity level. To add further complication, no improvement could be realized until the concentration of the product in the rich solvent stream had been reduced to trace concentrations. The transition to low product loss operating conditions took multiple weeks to achieve. All plant operating and business personnel had to be well informed on the length of time required to achieve the new operating conditions and that the critical impurity concentration in the product would be at or above the sales specification during the transition. It was also important to properly communicate the fact that none of the benefits would be realized until the very end of the transition. Operations and business personnel had to resist the urge to reduce the contact column temperature during any critical impurity excursions that took place during the transition period. They had to trust that by weathering the storm and achieving the low product loss conditions, excursions of high critical impurity concentrations would be eliminated.

After a multiple week transition period, the extractive distillation system was successfully transitioned to the new operating conditions with the following results:

- Product losses in the solvent system were virtually eliminated
- Critical impurity concentrations in the product were reduced to and maintained at levels significantly below product specifications
- All other impurities in the product were reduced to all time low concentrations
- The process achieved record operating rates.

Dividing Wall Column Revamp Optimises Mixed Xylenes Production

Bernie Slade and Berne Stober, ExxonMobil
bernie.slade@exxonmobil.com / berne.k.stober@exxonmobil.com

Dave Simpson, Koch-Glitsch, Stoke-on-Trent, UK
Dave.Simpson@KochGlitsch.com

Keywords: Distillation, Xylene, Dividing Wall, Energy Saving

Abstract

The successful revamp of a conventional tray Xylenes Distillation Column (3800 mm / 4300 mm diameter) is described. The column takes a xylene side-stream from reformate to feed an aromatics plant to make higher value products. Changing the column configuration to a Dividing Wall Column (DWC) resulted in improved product purity and significant energy savings. The project cycle is detailed including process design, hydraulic design, mechanical design, manufacture, testing and site work. Economic benefits are outlined and the lessons learned are discussed.

Introduction

ExxonMobil Fawley Refinery near Southampton in the south of England operates a distillation column (Figure 1) to recover mixed xylenes from reformate motor gasoline. The recovered mixed xylenes product is sent to an aromatics unit as feedstock for more valuable products. ExxonMobil has many ongoing projects to improve energy efficiency worldwide through technology development and implementation. The deployment of Dividing Wall Column (DWC) technology for the revamp of the Fawley xylenes column was identified as a good opportunity to achieve significant energy savings and improved product purity.

Figure 1. Xylene Column at ExxonMobil Fawley Refinery

DWC Technology

The concept of the DWC dates back to the 1940s and has been the subject of many papers. The first U.S. patent describing use of a vertical wall in a side-stream fractionation column, used to improve both the front end and back end fractionation of the side-stream, was by R. O. Wright.[1] Mr. Wright was an employee of the Standard Oil Development Company, a forerunner of the ExxonMobil Research and Engineering Company. The principles of the DWC will not be restated here but are well explained in the literature.[2,3,4,5,6,7]

DWC technology has been applied by European companies for the last 45 years. There are now approximately 70 towers in operation around the world, with the largest tower measuring several meters in diameter. Many of these towers use mass transfer equipment supplied by Koch-Glitsch. Until recently there has been relatively little commercial application of the full DWC concept at ExxonMobil.[8] However, there were applications of a partial dividing wall. For example, within ExxonMobil, internal vessel partitions have been used to segregate feeds for different disposition of bottoms products.[9] There was also a single instance of an internal side rectifier installed in an ExxonMobil linear paraffin unit tower. But within ExxonMobil there was no full DWC in operation until the Fawley xylenes tower application. Therefore, consistent with company procedures, the DWC technology went through a rigorous technical evaluation and risk mitigation before its implementation.

There is additional complexity involved in the design of a DWC. There are many more degrees of freedom available to the designer of a DWC compared to a conventional distillation train, which makes optimisation difficult and time consuming. A further restraint has been the perceived complexity and uncertainty of the control system. However, with the relatively cheap availability of computers and simulation packages, the design is far less daunting. In addition, the issues with the control system have been mitigated by modern computerised control systems.

The time available for revamp work might be one reason for avoiding a DWC. But as pointed out later in this paper, significant advances in this area have been made. So it is expected that more applications will occur especially as the value of energy conservation projects increase. The authors estimate that the design process is an order of magnitude more complex than a conventional route, but this is negligible when compared to the 30% savings in energy and also in reduced capital costs for a new plant.

Process Design

The existing column configuration (Figure 2) was a conventional column with 51 valve trays. The feed was liquid phase to tray 38. The mixed xylenes product was taken as a vapour side draw above tray 20. Above this elevation, because of the decreased vapour traffic, the column diameter was swaged down from 4300 mm to 3800 mm inside diameter. The column had an air-cooled condenser and a horizontal thermosyphon reboiler (33 barg steam). The xylene product was also cooled by an air-cooler.

The existing column performance was modelled to provide a basis for the DWC modelling and to establish accurate tray efficiencies for use in the DWC design. The modelling tool used was Provision.

Because the column size was fixed and it was desired to minimise the complexity of the revamp the total number of trays and, therefore, the number of theoretical stages were also fixed. Further, it was desired to re-use the existing reboiler and condenser. These constraints limited the number of degrees of freedom available, which simplified the process design. The choices to be made were the number of trays above the Dividing Wall (DW), the number of trays below the DW and the location of feed and product draw. The number of trays above and below the feed on the pre-fractionation side of the DW and the draw on the product side of the DW were determined by the locations of the feed and product draw, respectively. Finally, the liquid split at the top of the wall and the vapour split at the bottom of the wall had to be selected to set the L/V ratio in each section.

Initially the design did not look as though it could be achieved within the given constraints. However, it was determined that the xylenes product could be taken in the liquid phase, which reduced the vapour traffic and allowed a practical design. This also reduced the requirement for cooling the product.

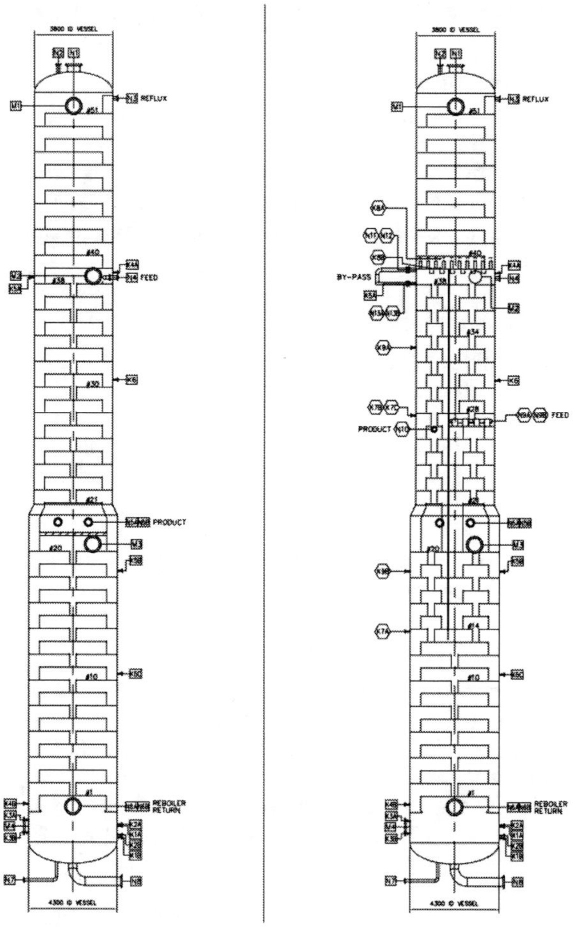

Figure 2.

Xylenes Column Pre-Revamp

Xylenes Column Post-Revamp

Economics dictate the optimum degree of desired xylenes recovery, which leads to a wide range of operating conditions for the column. Two quite different design cases and all of the conditions in between had to be considered.

In addition to the design case simulations, a number of sensitivity studies were required. The effects of heat transfer and leakage across the wall had to be considered. The former to determine if insulation of the wall was required, and the latter to determine the degree of seal required for the wall. The effect of errors in the estimated tray efficiency had to be determined. The tray bubble area shape is unusual and it was not clear that the empirical methods used by ExxonMobil would therefore accurately predict the efficiency. And finally and most importantly, the effect of the liquid and vapour splits had to be studied. From these studies, it was determined that (1) insulation was not required, (2) the seal of the wall was most critical around the feed and product draw, (3) it would be necessary to vary the liquid split at the top of the wall to operate the column at all desired conditions, and (4) reasonable errors in the assumed tray efficiency will not jeopardize performance.

One finding that was critical to both the process design and the hardware design and the final risk assessment of the technology was that errors in predicting the vapour split could be substantially corrected by changing the liquid split. Even though the tower was being designed with sieve trays, which have over 6,900 pressure drop data points available to members of Fractionation Research, Inc., there are still issues regarding the prediction of pressure drop. In the DWC, the pressure drop prediction is even more critical than a normal tower, because the vapour split is determined so that the differential pressure drop across the wall section is zero. Simulation work was done to investigate the effect of arbitrary errors in the vapour split, with control of the liquid split and without control of the liquid split. The results (Figure 3) show that control of the liquid at the top of the wall enables correction of any vapour split error.

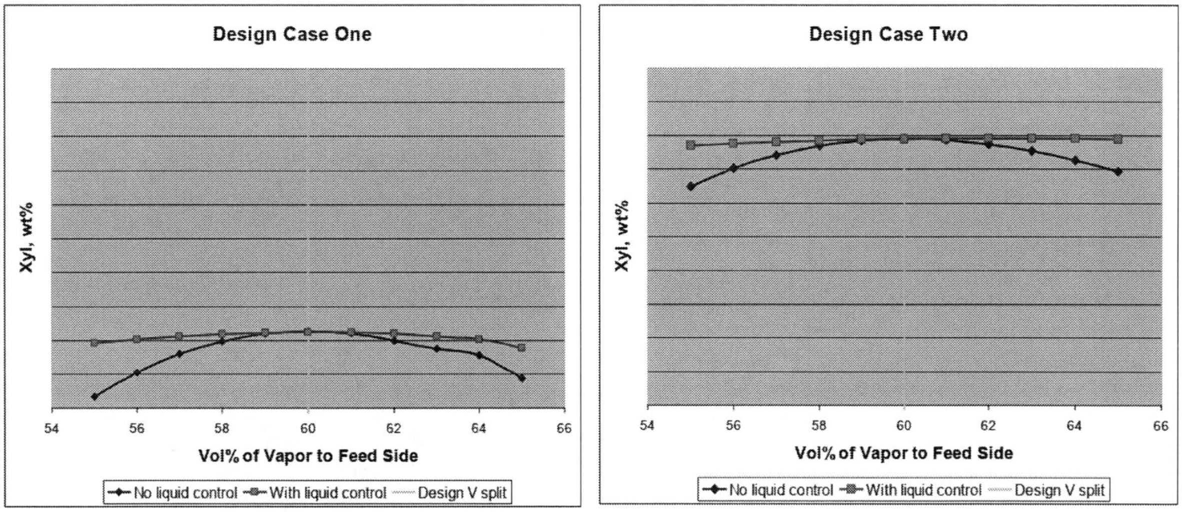

Figure 3. Product Purity Sensitivity to Vapor Split with and without Liquid Split Control

The revamp column configuration (Figure 2) was a DWC with 51 trays. The DW ran from tray 14 up to tray 39. The feed was two phase to tray 27, which was also a chimney tray. The xylenes product was taken as a liquid side draw from tray 28, which was an active tray. A liquid splitter tray was installed at tray 39. The column condenser and reboiler did not require any modifications. The product cooler was also re-used without modification although it is now rather oversized.

The key parameters of the column performance before and after the revamp are listed in Table 1. They are compared to the Pre-DWC operation as percentages.

Parameter	Pre-DWC	Case 1	Case 2
Operating Pressure	0.5 barg	0.5 barg	0.5 barg
Feedrate	100%	100%	103%
Xylenes Purity	100%	100%	106%
Xylenes Recovery	100%	100%	100%
Reboiler Steam	100%	47%	74%

Table 1. Operational Parameters

At case 1 operation the benefit is an energy saving of over 50%. Case 2 operation allows higher feed rate and product purity with an energy saving of over 25%. The fact that the DWC allowed to take a liquid phase xylenes product draw contributed to the energy savings.

Hydraulic Design

The column was amply sized for the required duties both before and after the revamp, thus the hydraulic design was not challenging with respect to capacity. The trays were rated at ~ 80% of capacity (limited by jet flood) at the maximum loads. It was not necessary to use high capacity trays or reduced tray spacing to obtain more theoretical stages[10]. Due to the change from vapour to liquid product draw the column is now rather oversized below the swage.

The primary concern in the hydraulic design was the operability of the column. This meant that the trays had to operate efficiently over a reasonably wide range of conditions. To this end, different designs were optimised for each section of the column. Sieve trays were selected for because their pressure drop has been extensively studied and because capacity was not limiting.

It was important to get the correct vapour flow rate split on both sides of the DW. This was achieved by balancing the pressure drop through the DW section at the required flow rates on either side of the wall. Design features that are used to optimise this balance included offsetting the DW from the column centre line between the feed and product draw sides. This also helps balance the flood values on either side of the wall. The open area of the trays was varied in each section and weir blocks were used to avoid potential spray regime operation. The pressure drop was balanced very closely at the key design load. Inevitably at off-design conditions, the pressure drop balance will vary, which results in a change in the vapour split. At the other extreme of the design conditions, a shift in the vapour split was calculated. This was examined through a process simulation sensitivity study and found to be acceptable, provided that liquid split control is used to compensate.

The liquid split at the top of the wall is just as important as the vapour split. If a constant split is required, this can be achieved with a chimney tray to meter the liquid on either side of the wall in a fixed ratio. In this case, the process design required a variable split. One solution was to draw off all of the liquid and pump it back while controlling the flow to either side of the wall. This method requires a quite complicated and expensive system. To avoid this expense, a novel hybrid was developed using a chimney tray to meter the bulk of the liquid in a fixed proportion and a gravity fed external by-pass to control the liquid split as required. This novel arrangement was tested in a full-scale test rig at Koch-Glitsch's Kirkby Stephen works (Figure 4). The test not only validated the design and calibrated the by-pass but allowed several improvements to be made and future simplifications to the design.

Figure 4. Liquid Splitter Test Rig at Koch-Glitsch

The secondary concern in the hydraulic design was to make the revamp as simple as possible. The existing two-pass arrangement was retained re-using the existing downcomer chords. This meant virtually no modifications were needed for the existing tower attachments. The DW was designed perpendicular to the downcomers, which again meant no modifications to the existing tower attachments and that the orientations suited the existing nozzles and trays above and below the DW. The existing trays above and below the DW were checked for the new duties and reused with only minor modifications (blanking).

Mechanical Design

The key concerns for the mechanical design were safe installation and maintenance, speed of installation to minimise column downtime, and economy in manufacture.

The trays in the DW section were two-pass sieve trays designed to fit in either side of the DW. To speed the tray installation, FLEXILOCK® tray construction was used (Figure 5). These boltless panel joints allow a 30% saving in installation time compared to traditional bolted joints. The trays also had boxed downcomers at the DW end to simplify attachment to the DW.

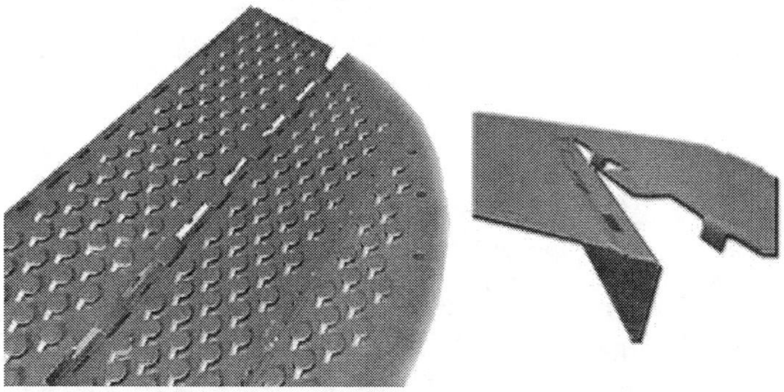

Figure 5 - FLEXILOCK® Tray Construction

The design of the DW (Figure 6) was the most important consideration. It was ~ 28 m tall and weighed ~ 4,000 kg. It had to be installed as quickly as practical, provide a good seal between the feed and product draw sides and support the trays. The design used new bolt-bars welded on both sides of the column. Cruciform sections were welded between the bolt-bars to provide the required structural strength and supports for the trays. The wall itself consisted of panels bolted between the bolt-bars and the cruciform sections. It was sealed with PTFE gasket except at the feed and draw-off locations, which were seal, welded. Some panels were designed to be easily removable as manway sections to aid future maintenance by improving access through the column.

The design required several new nozzles to be installed for the by-pass, feed and product draw-off plus new instrumentation. These were originally specified as set through; however, changing them to set on allowed them to be installed with minimum disruption to the work inside the column thus speeding the installation schedule.

Figure 6. Trial Assembly of DW at Koch-Glitsch

Installation

The installation was performed safely and within schedule by Koch-Glitsch in 30 days during May 2005. Because the xylenes column was the only part of the plant shutdown, the installation was surrounded by live plant. Special precautions had to be taken during the hot work, particularly outside the column. Meticulous planning and superb cooperation between the ExxonMobil plant personnel and the Koch-Glitsch site crew were responsible for the successful installation.

Operation

The existing conventional energy balance control was retained with the temperature control point relocated to below the DW.

Test runs were carried out on the DWC during June - July 2005 proving its performance across the complete design range and beyond. No particular operability issues were discovered during the test run.

Conclusion

Lessons learned about the design of the liquid splitter chimney tray will allow more economical designs in the future. Very important lessons were learned about the design and installation of the DW. Changes to the construction and installation sequence are estimated to reduce the installation time by up to 50% with consequent savings in site costs and downtime.

ExxonMobil has classified this revamp as a complete success and within ExxonMobil, DWC technology is now fully endorsed and is being incorporated in additional projects. In the Fawley tower, energy savings of up to 53% have been achieved together with improvements in xylenes purity. Similar improvements in energy efficiency are expected in other applications. The Fawley project payback was very healthy. The Dividing Wall column technology has several potential future applications at the manufacturing sites and is expected to be a significant factor in ExxonMobil's continuing drive to reduce energy consumption at its refineries and chemical plants.

Authors

Bernie Slade, Fractionation Specialist, ExxonMobil Research and Engineering Company

Bernie Slade is a fractionation specialist with ExxonMobil Europe Ltd., based in Fawley UK. He provides a wide range of fractionation services to ExxonMobil's refineries and chemical plants in Europe and the Middle East, ranging from planning, design, detailed engineering, commissioning, start up, optimisation and troubleshooting.

Berne Stober, Fractionation Specialist, ExxonMobil Research and Engineering Company

Berne Stober is the Fractionation Technology Group head at ExxonMobil's Central Engineering Office located in Fairfax, VA. His group is responsible for fractionation technology for ExxonMobil's worldwide downstream operations. One of his group's responsibilities is the vetting of all new fractionation technology implementations. In addition, the group provides fractionation services to the entire downstream ranging from planning, design, training & tools, commissioning, start up, optimisation and troubleshooting.

Dave Simpson, Mass Transfer Business Manager, Koch-Glitsch

Dave Simpson has been with Koch-Glitsch since 1989 and has vast experience in the design of a wide range of column internals for mass transfer. He is responsible for a team of engineers providing process design and sales from the UK office.

FLEXILOCK® and KOCH-GLITSCH® are registered trademarks of Koch-Glitsch.

References

[1] Wright, R. O., US Patent 2471134, Assigned to Standard Oil Development Company, May 1949.
[2] Lestak, F et al., "Advanced Distillation Saves Energy & Capital" *Chemical Engineering*, July 1997, pages 72-76.
[3] Ennenbach, F. et al., "Divided–Wall Columns – a novel distillation concept." *Petroleum Technology Quarterly*, Autumn 2000, pages 97-103.
[4] Schultz, M.A. et al., "Reduce Costs with Dividing-Wall Columns." *Chemical Engineering Progress*, May 2002, pages 64-71.
[5] Becker, H. et al., "Partitioned Distillation Columns – Why, When & How." *Chemical Engineering*, January 2001, pages 71-74.
[6] Becker, H. et al., "The World's Largest Partitioned Column with Trays – Experiences from Conceptual Development to Successful Start-Up." *Linde Reports on Science and Technology, 62/2000*, pages 42-48.
[7] Lestak, F. et al., "Heat Transfer Across the Wall of Dividing Wall Columns." *Trans. IChemE, Vol 72, Part A*, September 1994, pages 639-644.
[8] Parkinson, G., "Drip and Drop in Column Internals." *Chemical Engineering*, July 2000, pages 27-31.
[9] Graven, R. G., US Patent 3412016, Assigned to Mobil Oil Corporation, Nov. 19, 1968.
[10] Spencer, G. et al., "Consider dividing wall distillation to separate solvents." *Hydrocarbon Processing*, July 2005, pages 91-94.

Investigation on a Novel Double-effective Packing Element

Friederike Kutzer and Johann Stichlmair, TU München, Garching/München, Germany

Introduction

The technical term maldistribution denotes all deviations from plug flow of gas and liquid in separation columns. Especially in packed columns maldistribution is a serious problem since its degree as well as its negative effect on separation efficiency are very high [1]. In industrial practice the problem of maldistribution is counteracted by dividing the packed bed into several sections of limited height (e.g. up to 6 m). Between all these sections the liquid is collected, remixed and uniformly distributed to the bed beneath. In large columns, the entire redistribution device can reach a height of up to 2 or 2.5 m [2] . However, these very complex and expensive redistribution devices do not directly contribute to the interfacial mass transfer in the column. Thus, a novel packing element that provides both flow redistribution and mass transfer is developed with the aim of replacing the redistribution sections. The design of the novel double effective packing element is described in detail in [3]. Four promising installation structures of the novel packing elements are investigated below. To obtain an impression of the liquid stream profile inside the packing the maldistribution factor MF_L is plotted against the column height. This allows the determination of the optimised installation distances of the packing elements.

Novel packing element

The novel packing element, shown in Fig. 1, consists of a structured packing with an alternating arrangement of plane and corrugated sheets (see Fig. 1 a). This structure forms a system of separated channels with countercurrent flow of gas and liquid. The alternating inclinations (angle 45°) in parallel layers provide strong radial distribution of gas and liquid. Furthermore, a local liquid stream on the top of the packing element is split into several partial streams which flow downwards in varying directions. Thus, a high level of cross-mixing is provided too. To avoid increased wall flow of the liquid, the plane sheets are provided with a special construction of openings (1) and flaps (2) [4] at the lateral sides near the column wall (see Fig. 1 b). The flaps seal the channels and steer the liquid streams through the openings into the channels of the adjacent layers with different inclination. Thus, the liquid streams are recirculated into the packing and wall flow is effectively suppressed.

To gain knowledge of the operation characteristics of the novel packing element, experiments on a column totally filled with the novel packing elements are carried out (column height 3.72 m). These investigations provide data on pressure drop and separation efficiency. Because of the lower mass transfer area the novel packing element obtains approximately half of the separation efficiency of a conventional packing.

Several pairs of such elements, each approx. 20 cm high, are directly installed into a conventional packed bed with an offset of 90° to one another.

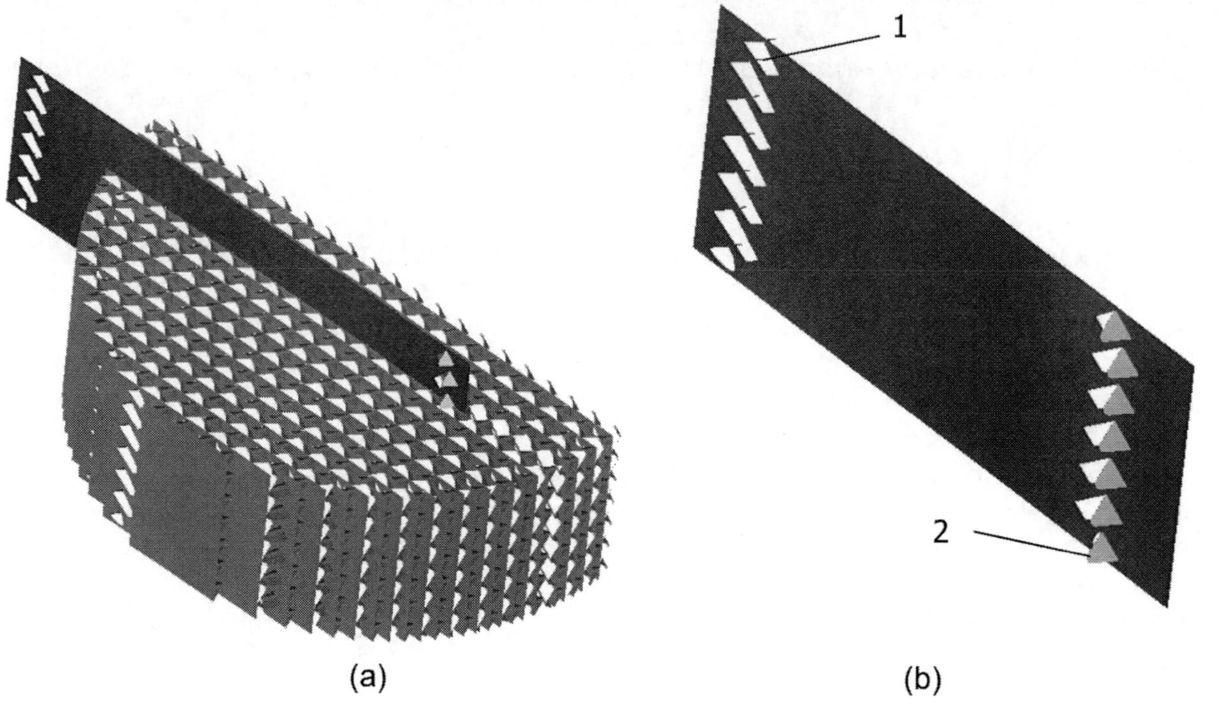

Figure 1. 3-D view of the novel packing element: general view (a); plane sheet (b) with openings (1) and flaps (2)

Experimental device and measuring technique

An established measuring technique was used to investigate the functionality of the novel packing element. In contrast to other known techniques, this measuring technique enables the visualization of the flow pattern inside the column. The technique is based on the determination of the liquid temperature profiles inside a packed column. For this purpose the column has to be operated with warm water and air like a cooling tower. The mass transfer of water from liquid to the vapour causes a significant temperature decrease of the liquid phase. This temperature variation can be easily measured and evaluated.

The test column used for the present experiments has a diameter of 0.63 m and a maximum packing height of 6.82 m. 732 thermocouples are evenly distributed inside the packing on 12 horizontal measuring cross-sections. A detailed description of the experimental device in combination with a schematic flow diagram can be found in [5].

The visualization of the liquid temperature profiles is performed by means of isotherms plotted in vertical cross sections of the column (see Fig. 2-4). In the case of plug flow, these isotherms are supposed to be straight horizontal lines. Any deviations from straight lines are caused by maldistribution. A first approximation considers the isotherms as lines of constant residence time. Lower regions of an isotherm mark areas with higher liquid load. In this way, the temperature profiles can be correlated with flow patterns of the liquid.

The extent of maldistribution strongly depends on the initial liquid distribution and the liquid distribution performance of the packing. For this reason the measurements were conducted with varying initial liquid distribution (uniform, central, annular and halfside) and two different kinds of random packings which highly differ in their spreading coefficient. The packings used were the Raflux-Ring 35-5 made from metal (modified Pall-Ring) which gives very good radial distribution, and the Hiflow-Ring 25-7 made from plastic, which is a representative of the group of grid packing with very poor radial distribution. Both the separation efficiency and the temperature profile within the conventional packing were determined. The first experiments were performed on a column solely filled with the conventional dumped packing mentioned above. These results serve as a reference for the experiments on a column additionally equipped with layers of novel packing elements. In order to optimise the distances and the number of packing elements, four different installation variations were investigated additional to the reference measurement.

Experimental results

Uniform initial liquid distribution

For uniform initial liquid distribution, Figure 2 compares the temperature profiles of the reference measurement (a) and the measurement with one pair of packing elements in the middle of the column (b). The isotherms are shown at 2°C intervals. The packing of the column consists of dumped Hiflow rings with a nominal size of 25 mm. The gas load *F* is 1.1 $Pa^{0,5}$, the liquid load B is 5.1 $m^3/(m^2h)$.

For the reference measurement, a steep inclination of the isotherms in regions near the column wall can be seen. This indicates that the liquid tends to distinctly develop wall flow. This distinctive wall flow amplifies down the packing. Compared to this, the picture on the right hand side shows impressively the positive effect of the novel packing element on the liquid stream profile. By installing one pair of packing elements the wall flow is effectively suppressed. This is reflected by the horizontal orientation of the isotherm beneath the packing element. Accordingly, a noticeable increase of the separation efficiency can even be achieved for uniform initial liquid feeding. The value for the height of a transfer unit decreases from H_{OG} = 0.85 m with the reference measurement to H_{OG} = 0.70 m with one pair of the novel packing elements installed in the dumped packing.

Halfside initial liquid distribution

Fig. 3 shows the liquid temperature profiles for halfside initial liquid distribution. The reference measurement on the left hand side (a) is displayed alongside the measurement with two pairs of novel packing elements (b). The first pair is installed near the top of the column, the second one is located in the middle of the column. The reference measurement provides a clear illustration of the poor self distribution capability of a standard dumped packing. The almost vertical orientation of the isotherms indicates that the packing is not able to noticeable improve the initial non-uniform liquid distribution. The liquid comes down the irrigated half of the column unchanged. The reason for this behaviour is related to the open grid structure of the Hiflow-Ring. Accordingly, the mass transfer is very unsatisfactory. This is expressed by a value of H_{OG} = 4.21 m for the height of a transfer unit.

Because of the inclination of the channels the novel packing element is able to steer liquid from the irrigated to the dry parts of the packing. The extent of the compensation of maldistribution depends (especially with halfside initial distribution) on the combination of column diameter, inclination angle of the channels and height of the packing element. A distinct improvement of the stream profile can be recognised from the orientation of the isotherms beneath the first pair of packing elements. But it is obvious that under the existing conditions the liquid is not distributed over the whole cross-section. With a maldistribution of this extent the insertion of a second pair of packing elements is reasonable. Even though the maldistribution is not completely abolished, the positive effect of the novel packing element on the stream profile is significant. This applies correspondingly to the effect on the separation efficiency. The height of a transfer unit can be reduced to H_{OG} = 1.37 m, i.e. it is improved by a factor of 3.

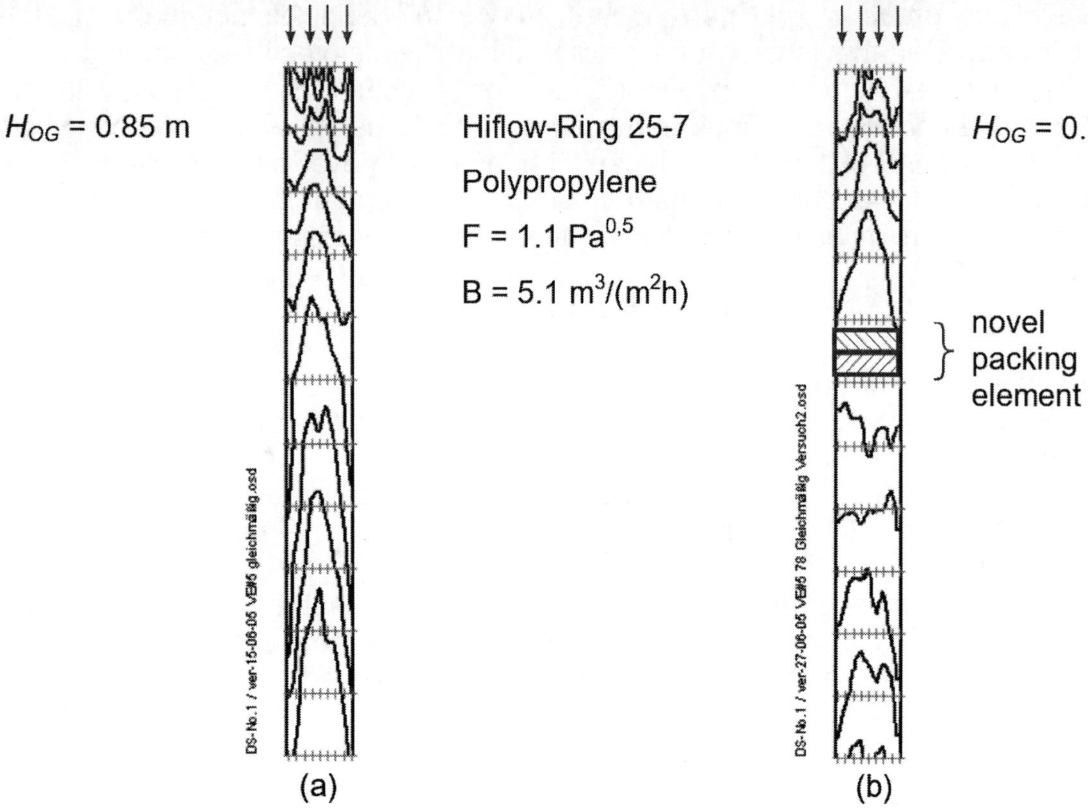

Figure 2. Liquid isotherms within a packing of Hiflow rings and uniform initial liquid distribution. (a) reference measurement with standard packing. (b) measurement with one pair of novel packing elements.

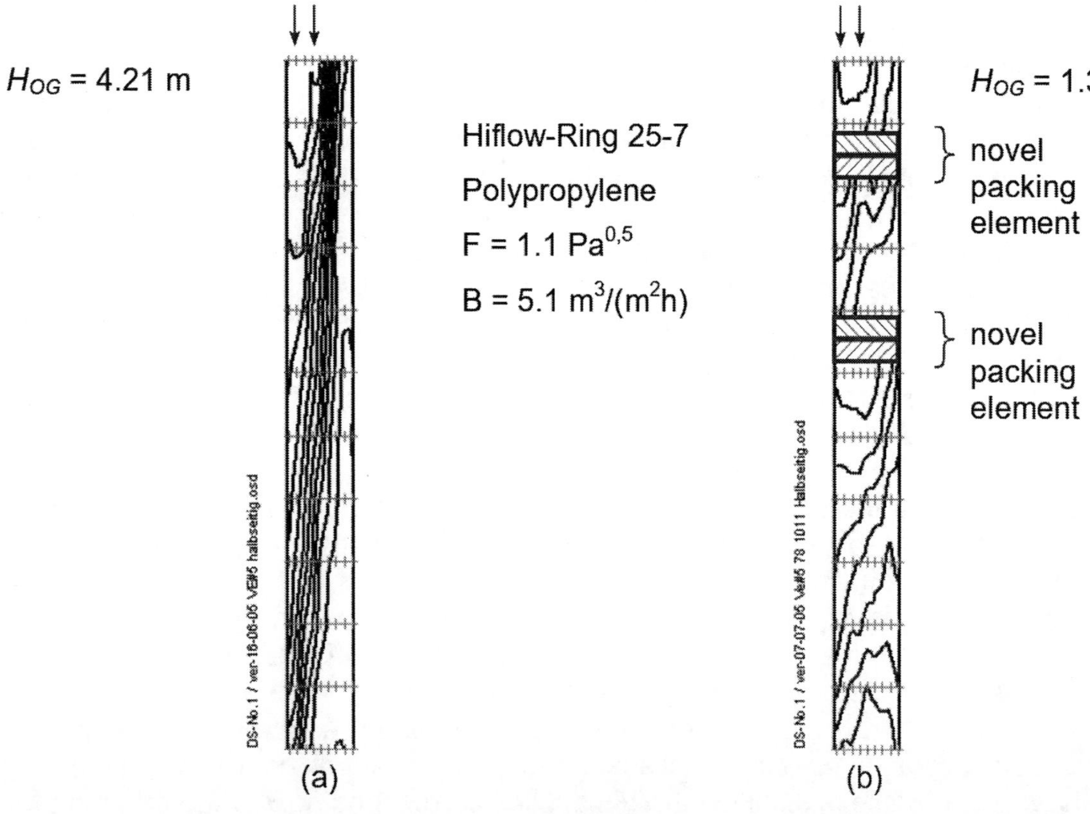

Figure 3. Liquid isotherms within a packing of Hiflow rings and halfside initial liquid distribution. (a) reference measurement with standard packing. (b) measurement with two pairs of novel packing elements.

Annular initial liquid distribution

By providing an annular initial liquid distribution, an existing wall flow ought to be simulated. The column on the left hand side (Fig. 4a) is fully stacked with the dumped packing. Here the isotherms maintain their steep slopes towards the column wall over the whole packing height. The packing is not able to correct the initial non-uniform liquid distribution on a large scale. The measured distribution after the whole liquid path of 6,82 m is far away from natural flow. The liquid flows down almost completely along the column wall. By inserting one pair of novel packing elements (see Fig. 4b), the wall flow can be suppressed effectively. The isotherm straight beneath the first pair of elements adopts an almost horizontal orientation. From this point on the temperature profile is similar to the one with uniform initial distribution. The liquid tends to develop wall flow again that increases on the way down the packing, until it is again suppressed by the second pair of packing elements. The thereby obtained improvement of separation efficiency is quantifiable. The height of a transfer unit is reduced from H_{OG} = 1.35 m in the reference measurement to H_{OG} = 0.79 m with two pairs of novel packing elements.

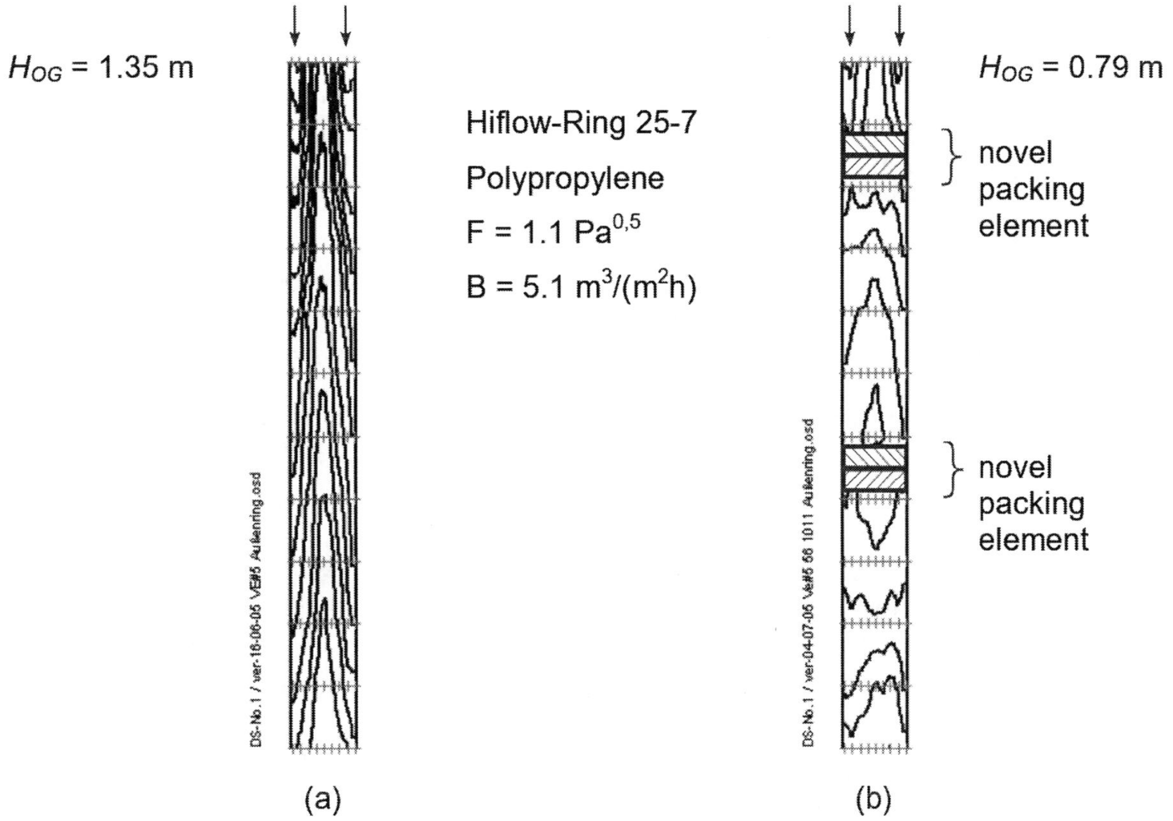

Figure 4. Liquid isotherms within a packing of Hiflow rings and annular initial liquid distribution. (a) reference measurement with standard packing. (b) measurement with two pairs of the novel packing element.

Evaluation of the experimental results

Separation efficiency

Fig. 5 shows an overview of the experimental results using Hiflow rings. The height of a transfer unit H_{OG} is plotted for the different kinds of initial liquid distribution for each installation variation of the novel packing element. For clarification a sketch of the particular installation variation is shown beneath the diagram.

For providing uniform liquid distribution it can be seen that the use of the novel packing element is not reasonable until a certain degree of maldistribution is compensated. This is related to the fact that the novel packing element itself achieves approximately half of the separation efficiency of a dumped packing. Thus with uniform initial liquid distribution the installation of one single pair of novel packing elements near the column head does not result in a distinctive improvement of the separation efficiency. But a pair of elements placed in the middle of the column leads to a noticeable improvement, i.e. a decrease of the height of a transfer unit H_{OG}. For the same reason, with uniform initial liquid distribution a second pair of packing elements is not required.

By contrast, with single point liquid feed an increase of the separation efficiency is only achieved through the use of a second pair of elements. Because of the geometric conditions at the test column (combination of column diameter, inclination angle of the channels and height of the packing element) the central liquid stream is steered directly to the column wall by the first pair. Thus there is just a displacement of the maldistribution from central to annular and the values for the height of a transfer unit remain at the level of the reference measurement. It is the second pair that is able to compensate this now existing wall flow. An improvement of separation efficiency from H_{OG} = 1.32 m to H_{OG} = 0.80 m can be achieved.

With an initial liquid distribution that strongly favours wall flow, such as annular distribution, the potential of the novel packing element can be demonstrated. This sort of uneven liquid distribution that often occurs in technical applications can be compensated with just one pair of novel packing elements. The comparison of the annular distribution values with uniform initial distribution values proves this. Both measurements obtain nearly identical values for the height of a transfer unit when a single pair of elements is installed near the column head. Using a second pair of novel packing elements the height of a transfer unit decreases from H_{OG} = 1.35 m to H_{OG} = 0.77 m.

The fact that dumped packing has a poor self-distributing tendency is illustrated by the result of the reference measurement with halfside initial liquid distribution. The value for the height of a transfer unit H_{OG} of 4.21 expresses the poor mass transfer of the column. Even though the liquid is not completely spread over the whole column cross-section by using one pair of the novel packing elements, the height of a transfer unit can be reduced to H_{OG} = 1.58 m. The best results are again obtained by installing a second pair of elements in the middle of the column. Values of H_{OG} = 1.35 m can be achieved.

Analogue to the experiments with Hiflow rings the experiments with Raflux rings are analysed. These results support the explanations given above for the results obtained with the Hiflow rings.

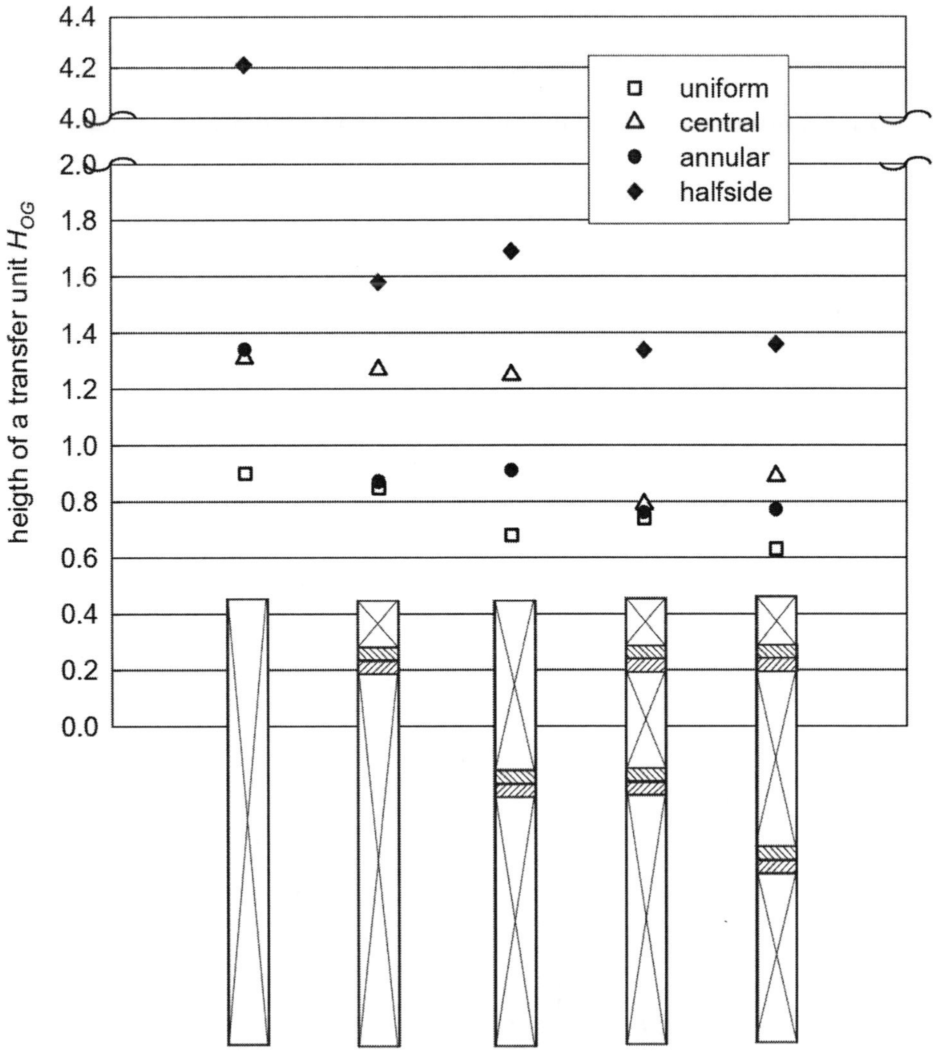

Figure 5. Overview of the separation efficiency with a packing of Hiflow rings and different installation variations of the novel packing element for different kinds of initial liquid distribution.

Stream profile inside the packing

In the literature the maldistribution factor MF_L [6], which is also known as the coefficient of variation C_v [7], is used to quantitatively characterize uneven liquid distribution. The maldistribution factor MF_L is defined as:

$$MF_L = \sqrt{\frac{1}{A}\sum_i A_i\left(1-\frac{v_i}{v_{av}}\right)^2} \qquad (1)$$

where A_i is the i-th area fraction of the column cross-section, A is the cross sectional area of the column, v_i represents the liquid velocity in the i-th area fraction of the column cross-section and v_{av} is the mean liquid velocity over the column cross-section. In the case of plug flow the value of the maldistribution factor MF_L is 0. The value increases with increasing extent of maldistribution. To get an impression of the stream profile characteristics **inside** the column, it is not sufficient to determine the maldistribution factor just for the column cross-section beneath the packing. From the results of the experiments on the test column the temperature profiles inside the packing are known. The decisive mechanism inside the column regarding maldistribution can be described by a model developed by Potthoff [8]. The local liquid streams are adjusted so that the calculated temperature profiles sufficiently match the measured temperature profiles. Once the local liquid streams of any column cross-section are known, the maldistribution factor can be determined from the equation above.

A result of this procedure is shown in Fig. 6. The temperature profile for the reference measurement with annular initial liquid distribution is shown on the right hand side of the diagram. This model result is compared to one with two pairs of novel packing elements. The maldistribution factor MF_L is plotted over the column height for both model results. Above the first pair of elements identical values for the maldistribution factor are obtained. With $MF_L = 0.78$ and $MF_L = 0.62$ the values are quite high due to the strongly non-uniform initial liquid distribution. That the novel packing element is able to suppress the default wall flow almost completely is reflected in the significant decrease of the maldistribution factor MF_L. Beneath the first pair of elements a value of $MF_L = 0.2$ can be achieved. Without the novel packing elements a value of $MF_L = 0.53$ is obtained. The tendency for the liquid to again develop wall flow and thus the increase of the maldistribution factor is avoided by installing a second pair of elements.

This diagram impressively shows the positive effect of the novel packing element on the liquid stream profile inside the column. By defining a maximum tolerable value for the maldistribution factor MF_L, optimised installation distances of the elements can be specified.

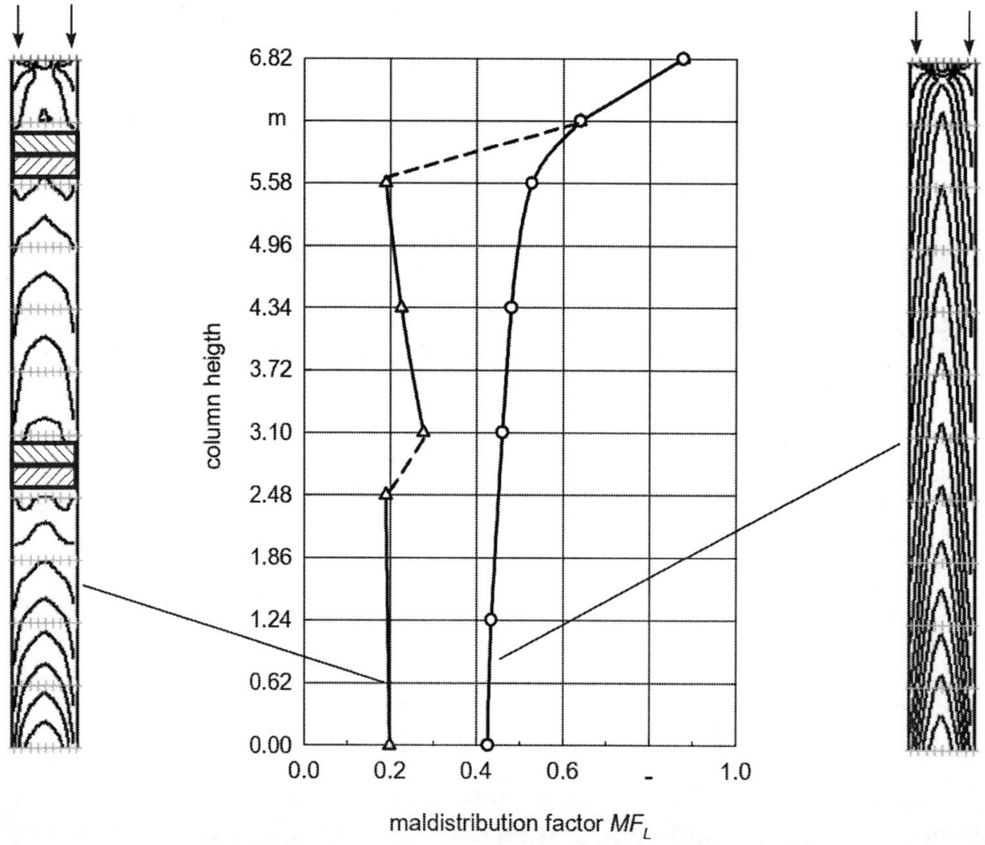

Figure 6. maldistributionfactor over the column height with annular initial liquid distribution. Comparison of reference measurement and measurement with two pairs of novel packing elements.

Conclusions

The implementation of a packing element that provides both flow redistribution and mass transfer is a promising approach for the replacement of conventional redistribution sections. Experimental investigations on a packed column, operated as a cooling tower, approve its positive effects on the liquid flow pattern and thus on the separation performance of the column. By using different kinds of random packings with varying installation variations of the novel packing element and several kinds of uneven liquid feeding the potential of the novel packing element can be demonstrated. For a quantitative characterisation of the liquid distribution inside the column the maldistributionfactor is determined for several column cross-sections. By plotting the maldistributionfactor against the column height, its course can be deduced to optimised installation distances for the novel packing element.

Nomenclature

A	cross sectional area, m^2
A_i	i-th area fraction of the column cross-section, m^2
v_{av}	mean liquid velocity over the column cross-section, ms^{-1}
v_i	liquid velocity in the i-th area fraction of the column cross-section, ms^{-1}
MF_L	maldistributionfactor
H_{OG}	height of a (vapour side) transfer unit, m

References

1 Stichlmair, J. G.; Fair, J. R.: *Distillation – Principles and Practice*. New York: Wiley-VCH 1998.

2 Kaibel, G.: European Patent No. 0 684 060 B1, 1995.

3 Schneider, O. and Stichlmair, J.: *Functionality of a Novel Double-Effective Packing Element*, Trans IchemE, Vol 81, Part A January 2003

4 Stichlmair, J. G.: European Patent No. 1 332 794 A2, 2003.

5 Schneider, O. and Stichlmair, J.: 3rd ECCE, Nürnberg (CD-ROM of full texts), 2001.

6 Kouri, R. J.; Sohlo, J.: *Liquid and Gas Flow Patterns in Random Packings*. Chem. Eng. J. 61 (1996), 95-105

7 Billingham, J. F.; Bonaquist, D. P.; Lockett, M. J.: *Characterization of the performance of Packed Distillation Column Liquid Distributors*. IChemE Symp. Ser. No. 142 (1997)

8 Potthoff, R.: *Maldistribution in Fülkörperkolonnen*. Fortschritt-Berichte VDI Reihe 3 Nr. 294. Düsseldorf: VDI-Verlag 1992.

88c Measurement of Packed Column Gas/Liquid Contact Areas for CO_2 Absorption

Frank Seibert, Separations Research Program, University of Texas at Austin, 10100 Burnet Road, Bldg 133/CEER, Austin, TX 78758 and J. Christopher Lewis, Separations Research Program, UT-Austin, 10100 Burnet Road, Bldg 133/CEER, Austin, TX 78758.
Measurements of packed column gas/liquid contact areas obtained using a variety of high performance random and structured packings will be presented. The effects of liquid and gas rates and surface tension will be discussed.

88d Debottlenecking an Acn Heads & Dry Column with Vg Af (V-Grid Anti-Fouling) Trays

Giuseppe Mosca[1], Elena Tacchini[1], Joshi Chandrakant[1], Shubhangi Jaguste[2], Prabhakar Nag[3], and Kanubhai Kalidas Parmar[4]. (1) Technology, Sulzer Chemtech, Piazza Duca D'Aosta, 12, Milano, 20124, Italy, (2) Technology, Reliance Industries, Mumbai, India, (3) Operation, IPCL, Vadodara, India, (4) Technology, IPCL, Vadodara, India

One of the major problems in many Refinery and Petrochemical Plants is the deposition of heavy organic solids dissolved in the process fluid or generated via polymerization of unstable components contained in the process fluid as well. Fouling is a leading cause of decreased capacity, excessive energy consumption, and or unexpected plant's shoutdown with consequent loss of production. Sulzer Chemtech has developped a new type of fractionation tray especially suitable for fouling-prone applications. The new VGAF (V-Grid Anti Fouling) tray provides high fouling resistance, high hydraulic capacity, and high fractionation efficiency as well. A great contribution has come from the experience of Reliance-IPLC at their petrochemical complex in Vadodara, India, where a successful revamp project was implemented at the Heads & Dry column of the Acrylonitrile plant. The upgrading of the operating procedure and the use of the VGAF trays allowed doubling the run length of the unit, while maximizing the capacity and the ACN recovery as well.

Modeling the Minimum Entrainment on Sieve Trays

Shuzo Ohe, Ph.D. Professor, Science University of Tokyo

Department of Chemical Engineering, Graduate School of Engineering

Science University of Tokyo, 1-3 Kagurazaka, Shinjuku-ku, Tokyo 162-8601, Japan

ohe@ci.kagu.tus.ac.jp

ABSTRACT

Entrainment on a distillation tray affects efficiency, pressure drop and the capacity of the distillation columns. Entrainment on a sieve tray increases with the increase of vapor load at constant liquid load. However, at constant vapor load, entrainment behaves somewhat differently. It has a minimum value. This phenomenon was uncovered by systematic study on a wide range of trays and systems at FRI (Fractionation Research Incorporated). Many arguments have been given for this phenomenon. Of these, the dominant one seems to be that the minimum is caused by the transition of vapor liquid contact from spray to froth on the tray.

This study, however, argues from an entirely different point of view. An assumption is made that the longer the residence time of the liquid on the tray, the greater the chances of generation of entrainment, The reason is that the tray specification, especially the height of the outlet weir, affects the phenomenon of minimum entrainment. It was found that data released by FRI exhibited this new relationship between minimum entrainment and the newly defined minimum residence time at constant vapor rate and increasing liquid rate for sieve tray. The liquid load at which the entrainment becomes minimum at constant vapor load operation coincides with the liquid load at which the residence time of liquid on the tray becomes minimum. In the newly defined residence time, the definition of volume is the product of the bubbling area and the observed clear liquid height. The observed clear liquid height has a significant effect on the residence time on the tray. The clear liquid height as reported by FRI was obtained from a series of experiments for pressure drop study.

This afforded an opportunity to predict the minimum of entrainment from only the value of clear liquid height. This relation can be applied to predict the minimum entrainment, if the clear liquid height is known.

We have examined several correlation methods for determining clear liquid height. Among them, we focused on Bennett's correlation, because it is the most recent of those examined. Like most other correlations, Bennett's is also based on Francis's equation. However, Francis's correlation itself is not adequate, due to an anomaly in its derivation: namely, the value of the exponent. Francis's correlated between clear liquid height and liquid load using an exponent with a value of 2/3. However, in actual fact this fraction does not fit the observed data. On checking, we determined that the value should be greater than unity.

INTRODUCTION

Until entrainment was observed systematically at FRI, the existence of minimum entrainment on the change of liquid load at constant vapor load was still only speculation. There were various kinds of arguments. Various investigators had found that an increase in liquid rate either increases entrainment, decreases entrainment, or both. One explanation offered was that at very low liquid rates, the low liquid level on the tray offers very little resistance to the flow of vapor and results in high entrainment rates, and that as the liquid rate increases, the resistance to vapor flow increases and reduces the energy of the vapor available for entraining drops of liquid, thereby lowering the entrainment rate. Another proposed explanation suggested that further increases in liquid load continue to reduce the entrainment but increase the liquid level on the tray and reduce the distance a drop must travel to the tray above, and that at some point, these effects are balanced and further increase in liquid rate results in higher entrainment.

Another approach involves the mechanism of vapor-liquid contacting on the tray. In relation to entrainment, this approach presents two regimes of vapor-liquid contacting: spray, and froth. The minimum point of entrainment is thought to occur at the transition from the spray to the froth regime. Various correlations about the transition have been presented.

We discerned this new relationship between minimum entrainment and minimum residence time at constant vapor rate and increasing liquid rate for a sieve tray from data released to the public by FRI. We have made an assumption that the longer the residence time of the liquid on the tray, the greater the chances of generation of entrainment. The newly defined residence time is related to the liquid volume on the tray. The volume is the product of the bubbling area and the clear liquid height. Fig. 1 illustrates the concept of this relation. The clear liquid height is thought to increase when the liquid load increases. The liquid load at which the entrainment becomes minimum at constant vapor load constant operation coincides with the liquid load at which the residence time of the liquid on the tray becomes minimum. The residence time is defined as Eq. (1), based on the concept shown in Fig.1.

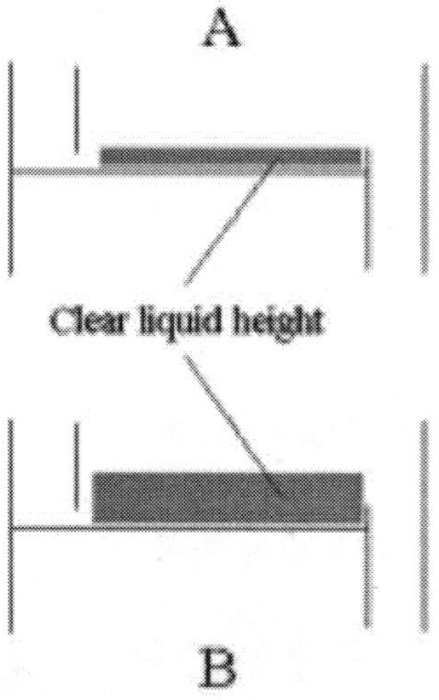

Fig. 1 Effect of clear liquid height on the entrainment

The clear liquid height shown in Fig. 1 is affected by the height of the outlet weir and the liquid load. For the same height of outlet weir, the clear liquid height will depend on the liquid load. In Fig. 1, in the lower case (A), the load is lower; and in the higher case (B), the load is higher. However, the residence time of the liquid on the tray is not reflected simply by the clear liquid height alone. Thus, minimum entrainment can result, depending on the change in the liquid load.

MINIMUM OF ENTRAINMENT AND RESIDENCE TIME OF LIQUID

The relationship between the observed clear liquid height against the vapor rate and outlet weir height is shown in Fig. 2 at constant liquid load. The clear liquid height decreases with increase in vapor rate. However, with an outlet weir height of 0, the clear liquid height is approximately 30 mm. In comparison, when the outlet weir height is 2 in., the clear liquid height is about 55 mm. The higher outlet weir height results in higher clear liquid height.

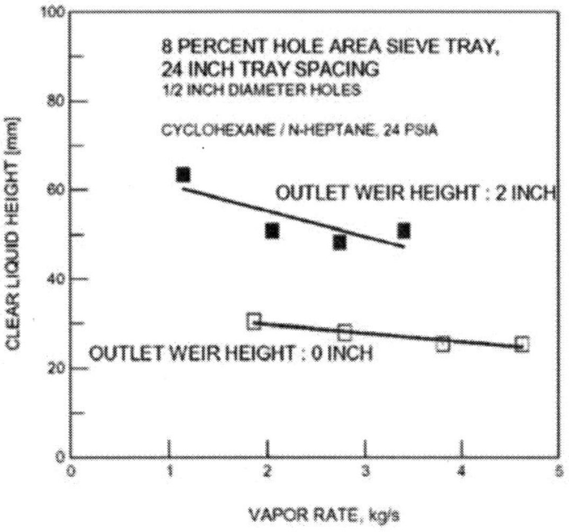

Fig. 2 Clear liquid height and outlet weir height

From experimental data on clear liquid height, the clear liquid height appears to be a function of outlet weir height under the same conditions of vapor and liquid loadings. This means that the residence time of liquid flow on the tray is affected by the height of the outlet weir. For the clear liquid height, some other observations were made as follows.

1. For low liquid rates, the clear liquid height on the tray is independent of the vapor rate. This appears to be a phenomenon of the spray regime.

2. For higher liquid rates, the clear liquid height on the tray increases with vapor rate. This appears to be a phenomenon of the froth regime.

3. Clear liquid heights on the tray do not appear to follow the Francis weir formula. Liquid heights measured at the highest liquid rates and high vapor appear to be greater than that calculated. This would imply that there is some other mechanism building up liquid on the tray. Possibilities are:

- Resistance to the flow of liquid into the downcomer due to vapor backflowing out of the downcomer.
- Vapor jetting close to the outlet weir, effectively shortening the outlet weir.

4. Other published methods do not appear to explain this high clear liquid level (Colwell, 1981).

From the above mentioned considerations, the residence time is newly defined as

$$\tau = \frac{v}{L} \quad (1)$$

where τ is the residence time [s], v is the volume defined as the product of the bubbling area Ab [m2] and the clear liquid height hcl [m],

$$v = A_b \cdot h_{cl} \quad (2)$$

and L is the liquid rate [m³/s] on the tray.

From Eqs. (1) and (2), it is clear that the volume v varies with the change in clear liquid height. Thus it also varies with the change in liquid loads.

Both terms in the fraction are functions of the liquid load. Therefore, the residence time does not change uniformly with the change in the liquid load. This is because the clear liquid height does not increase monotonically with the increase in liquid load. In other words, the clear liquid height value does not increase progressively in just one direction – i.e., not only up or only down.

RESULTS

The data observed by FRI, of entrainment tested with an 8% open area tray in a 4-ft tower, were examined. Figures 3 and 4 show entrainment data for cyclohexane/n-heptane at 24 psia and isopropanol/water at 15 psia. By cross plotting the data at constant liquid load, the minimum entrainment at constant vapor load is obtained as shown in Figs. 5 and 6. Figures 7 and 8 show the minimum of residence time determined by Eqs. (1) and (2), corresponding to the minimum of entrainments.

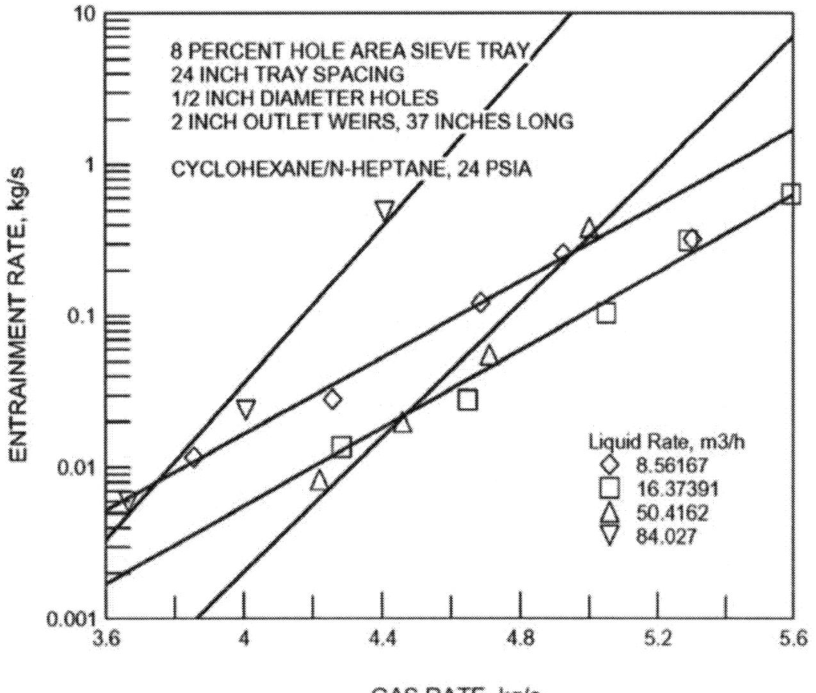

Fig. 3 Entrainment vs. vapor rate for cyclohexane/n-heptane system

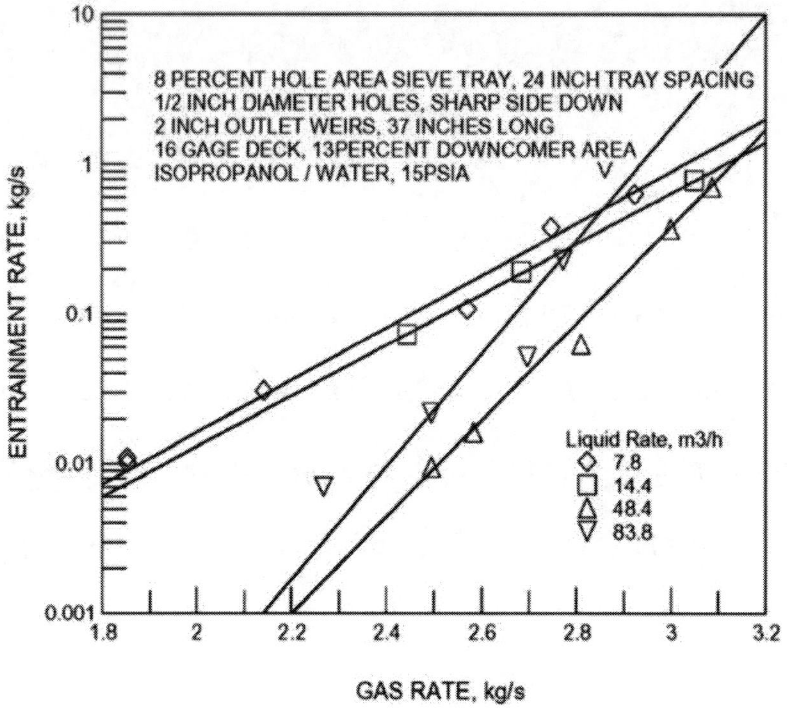

Fig. 4 Entrainment vs. vapor rate for isopropanol/water system

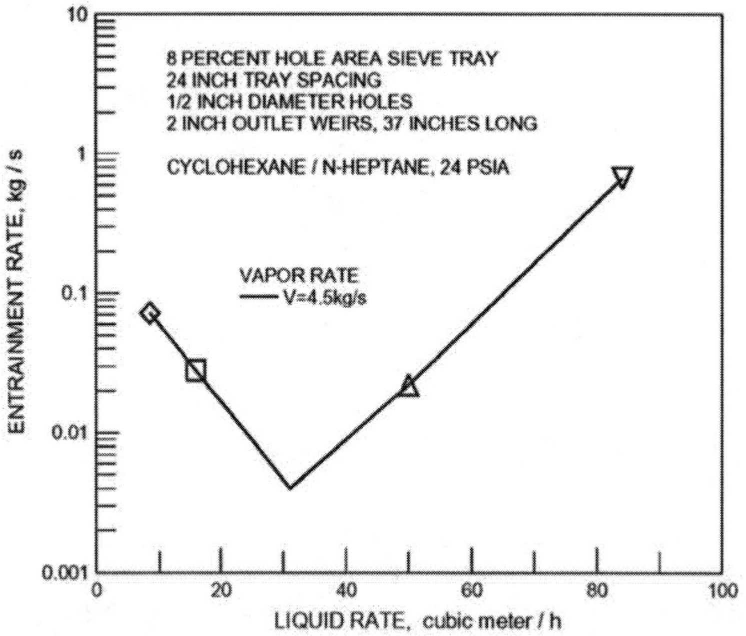

Fig. 5 Entrainment vs. liquid rate for cyclohexane/n-heptane system

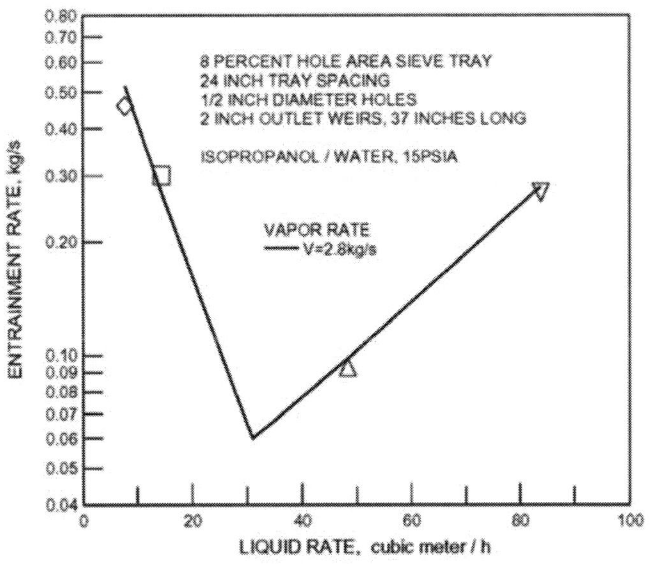

Fig. 6 Entrainment vs. liquid rate for isopropanol/water system

Figures 7 and 8 show the minimum residence time, as determined by Eqs. (1) and (2), corresponding to the minimum entrainment. In Figs. 5 through 8 it can be seen that the liquid rate at minimum entrainment and minimum residence time are the same at 30 m3/h, even though the systems and operating pressures are not the same. For the water/steam system at 3 psia, almost the same result was obtained. The liquid load at the minimum was 28 m^3/h.

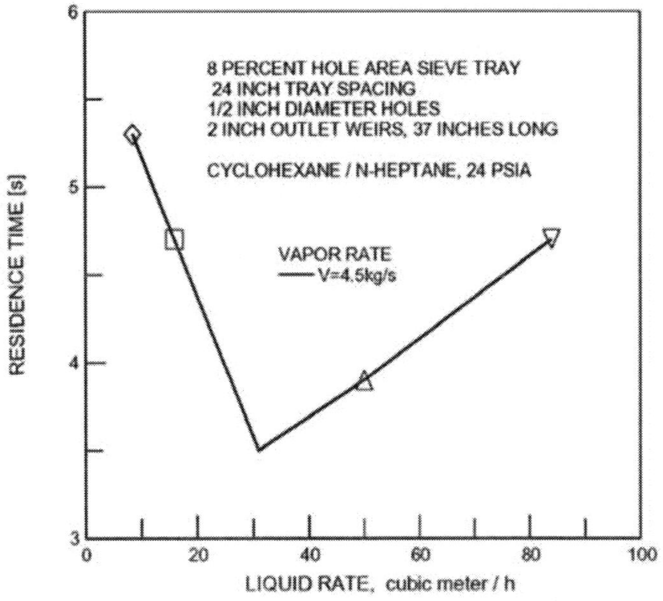

Fig. 7 Residence time vs. liquid load for cyclohexane/n-heptane system

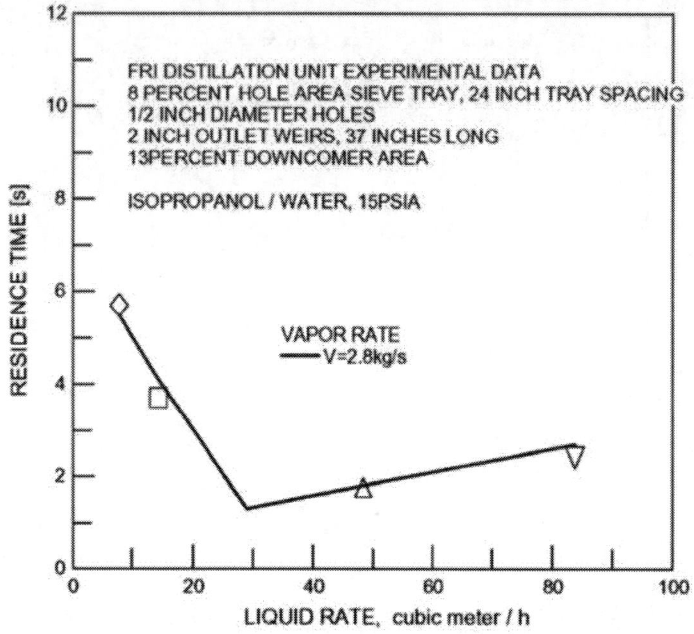

Fig. 8 Residence time vs. liquid load for isopropanol/water system

There are two kinds of coincidences. The first is that of the minimums of the entrainment and residence times. The second is that the minimum liquid loads are similar and the values are near 30 m^3/h. It may be that the first reflects the mechanism of the tray action and the second the tray geometry – mainly the outlet weir height.

Figure 9 shows the observed clear liquid height plotted against the vapor rate at constant liquid rates for the isopropanol/water system at 15 psia. At each vapor rate, the height increases with the increase in liquid rate. Figure 10 shows the residence time relationship [Eq. (1)], using the value in Fig. 9. Figure 9 shows that the clear liquid height increases with increases in liquid rate at each constant vapor rate. However, the residence time does not change in the same way as the clear liquid height. The residence time goes through minimum, as shown in the cross plot in Fig. 8. But notice that a change occurs in the sequence of liquid rate values. In Fig. 9, the highest liquid rate (83.8 m3/h) occurs at the highest clear liquid height, followed by the other liquid rates, in strict descending order. But in Fig. 10, the higher two rates are inverted; 83.8 m3/h is in third place and 48.4 m3/h is lowest.

Cross plotting the data in Fig. 9 reveals that the clear liquid height at minimum entrainment is about 21 mm. To determine the residence time accurately, the correct clear liquid height on the tray must first be predicted. However, the Francis relation for clear liquid height cannot predict the minimum residence time or clear liquid height. The actual geometry of the distillation tower is more complex than the Francis relationship can express. Additional parameters must be added, before the relationship can adequately describe these phenomena.

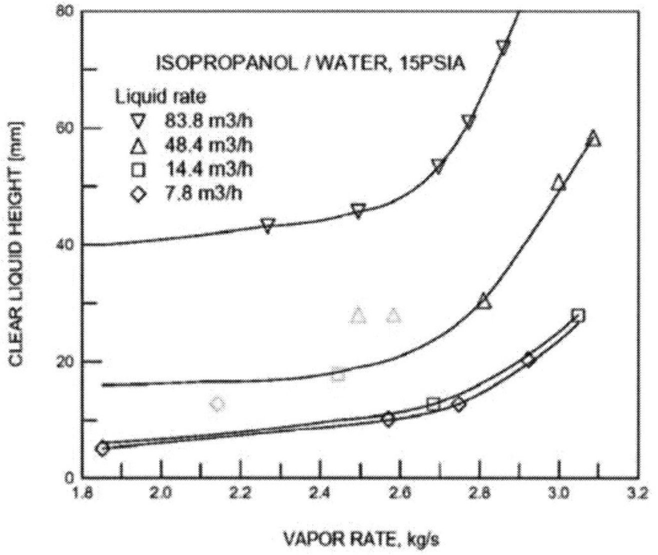

Fig. 9 Clear liquid height vs. vapor rate

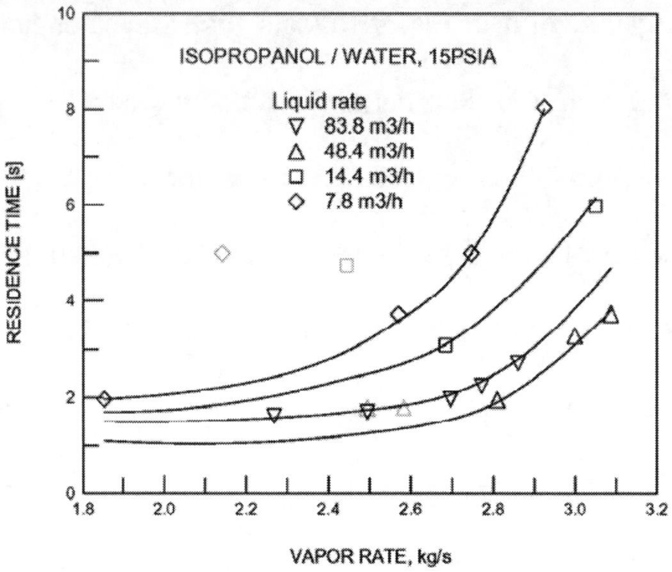

Fig. 10 Residence time vs. vapor rate

DISCUSSION

The clear liquid height plays a significant role in determining residence time. Equations have been developed that correlate clear liquid height with liquid load and other constituents. It is important to have a clear idea of just how closely these correlation equations could approximate the actual observable values. M. Locket's book presents three equations for clear liquid height:

Hofhuis & Zuiderweg (1979)

$$h_{cl} = (1-\varepsilon)\left[h_W + \frac{0.49}{C_d^{2/3}}\left(\frac{Q_L}{(1-\varepsilon_W)W}\right)^{2/3}\right] \tag{4}$$

Colwell (1981)

$$h_{cl} = \varepsilon h_W + 7.3\left(\frac{\varepsilon^{1/2} Q_L}{C_d W}\right) \tag{5}$$

Bennett (1983)

$$h_{cl} = \alpha_e\left[h_W + C\left(\frac{Q_L}{W\alpha_e}\right)^{2/3}\right] \tag{6}$$

The third, the Bennett equation, is the most recent one. Again, it can be seen that this equation shows the same influence of the two-thirds power upon the liquid load. These three equations are all based on the Francis equation. As a result, they each express the clear liquid height in the same way.

$$C = 0.50 + 0.438\exp(-137.8h_W) \quad (7)$$

$$\alpha_e = \exp\left(-12.55\left(V_B\left(\frac{\rho_V}{\rho_L - \rho_V}\right)^{0.5}\right)^{0.91}\right) \quad (8)$$

Figure 11 compares observed clear liquid height to that calculated using the Bennett equation. From this figure it is clear that the change in calculated clear liquid height is small. Also, the shapes of the two plots are different: the observed plot is a concave curve, but the calculated plot, while it may appear straight, in fact is slightly convex.

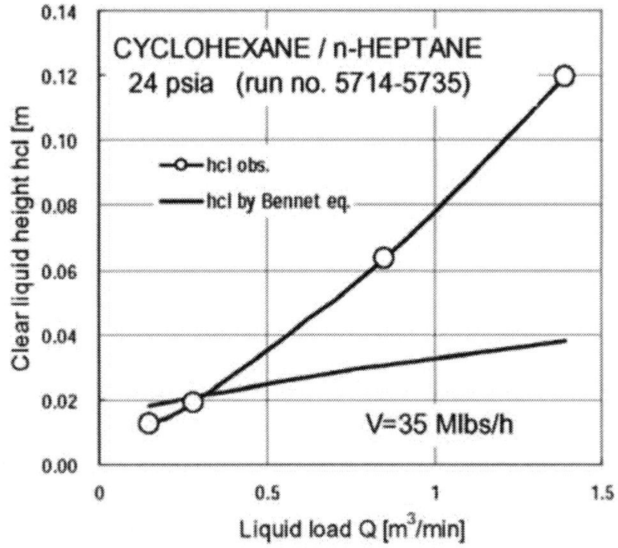

Fig. 11 Clear liquid height using Bennett correlation

Figure 12 shows the residence time, which is estimated using the observed and calculated clear liquid height values. From this figure, it is clear that the plot calculated using the correlation equation does not show a minimum.

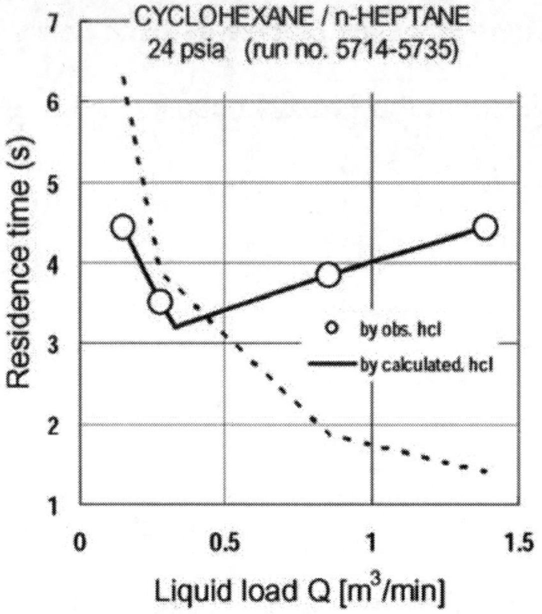

Fig. 12 Comparison of residence time

Figure 13 shows an enlargement of part of the graph presented in Fig. 11. The convexity of the calculated curve is even more evident. This is an effect of the term Q in the equation.

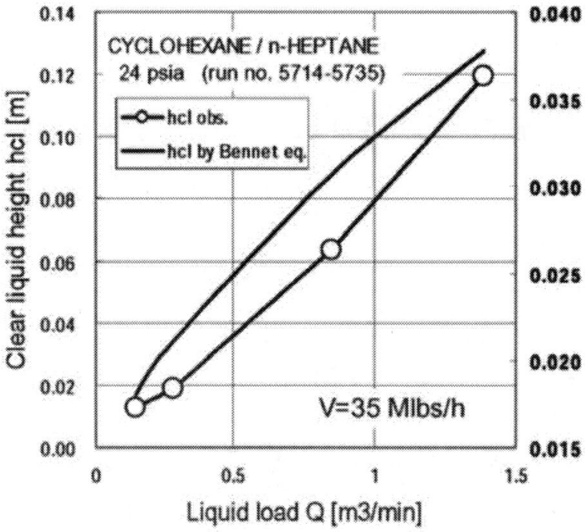

Fig. 13 Clear liquid height using Bennett correlation

Figure 14 is an example gained from the pressure drop study, not from the entrainment study. Nevertheless, the same result is obtained as from the entrainment study data. This figure shows the minimum residence time, as determined by Eqs. 1 and 2, that corresponds to the minimum entrainment at a liquid rate of about 110 gpm.

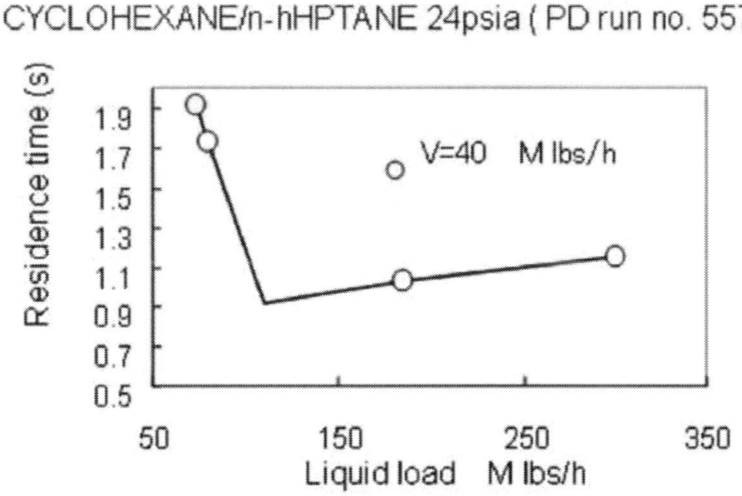

Fig.14 Minimum residence time from pressure drop study

Shown in Fig. 15 is the residence time, estimated using the observed and calculated clear liquid height values. It is clear that the plot calculated using the correlation equation does not show a minimum.

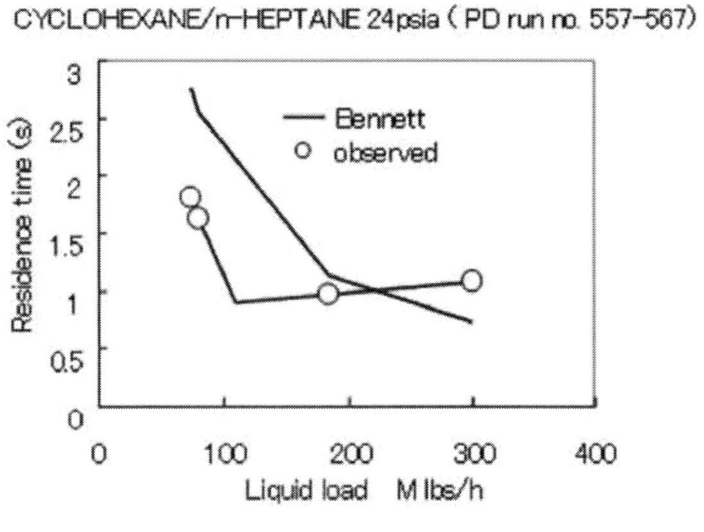

Fig. 15 Comparison of residence time

Figure 16 shows the clear liquid height, calculated using the Bennett correlation, of the case shown in Fig. 15. Again, the calculated curve is visibly convex. This is an effect of the term Q in the equation. In this figure, the curve plotted with circles shows the observed value, in a concave curve. The calculated value appears as a convex curve.

These results suggest that these correlation equations are inadequate to fully express the change of clear liquid height against liquid load. Consequently, a new way is needed to consider this correlation of clear liquid height with liquid load. The shape of the change in clear liquid height cannot be fully expressed by equations based on the Francis correlation.

It has been argued that, though it was theoretically derived, the two-thirds, or 0.67, power used in Francis-type equations is not the most accurate for the term Q. Thus, in the absence of highly reliable and realistic equations for calculating clear liquid height, the need is even greater for an accurate and easy-to-use empirical method.

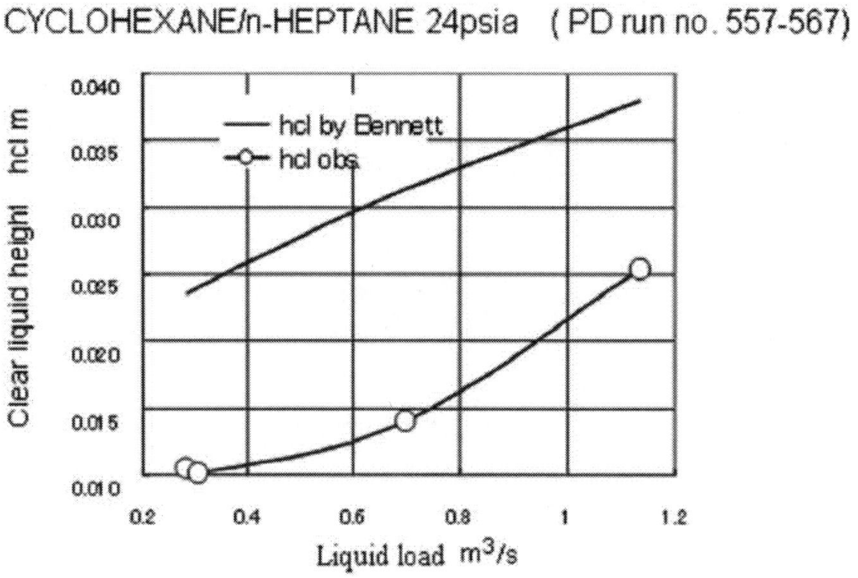

Fig. 16 Clear liquid height using Bennett correlation

Figures 17 through 19 show other examples. Each figure shows a similar phenomenon: the observed plots are concave, while the calculated plots are convex.

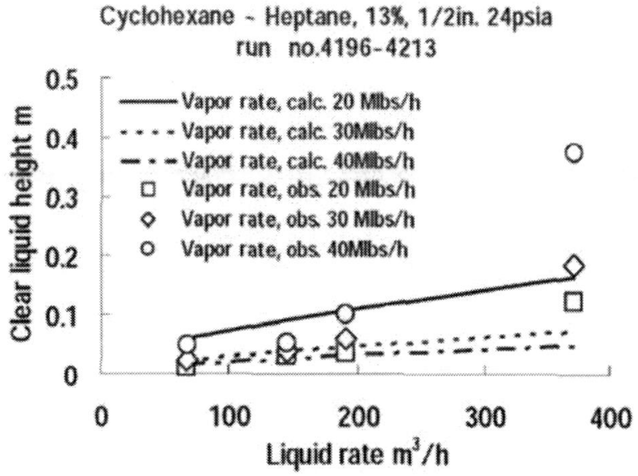

Fig. 17 Clear liquid height using Bennett correlation (1)

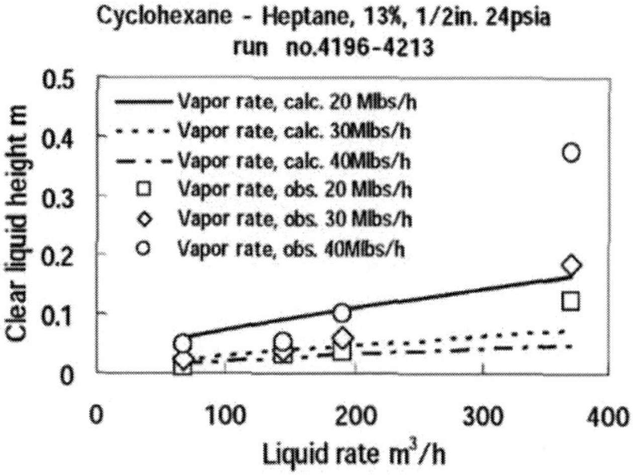

Fig. 18 Clear liquid height using Bennett correlation (2)

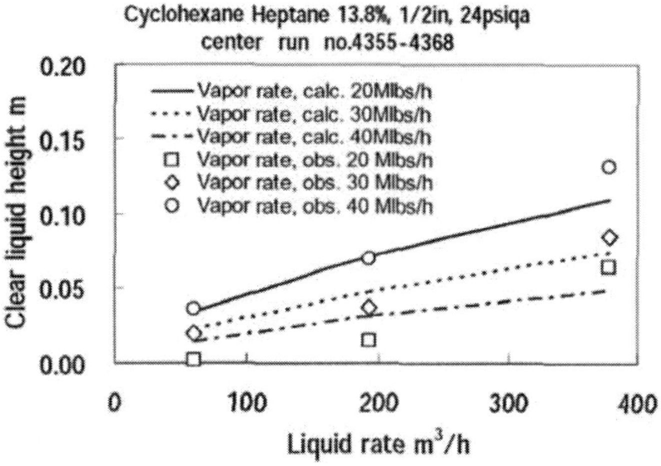

Fig. 19 Clear liquid height using Bennett correlation (3)

Figure 20 shows the relation between clear liquid height and residence time. The arrows indicate which Y axis to use – the left, or the right. The O group is the observed data and the B group is the result calculated using the Bennett equation. According to the Bennett calculation, the residence time continues to decrease. But as shown by the observed data, it reaches a minimum point and then becomes level – and in fact even increases slightly. So the Bennett equation does not express clear liquid height precisely.

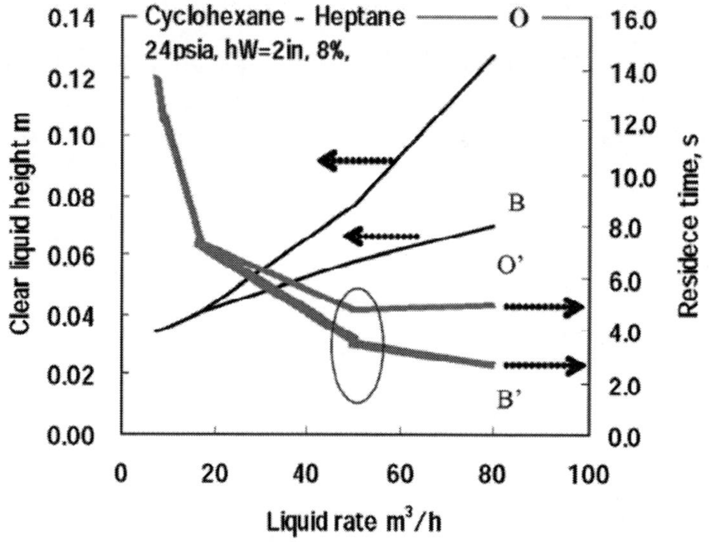

Fig. 20 Correlation of clear liquid height

Figure 21 shows the shape of the equation's function using five different exponents, from 0.5 to 2. The thicker black line is the result using the Francis-type power of 0.67. It is clear that when the exponent is 0.67, the line's shape is convex, but when the exponent is greater than unity, as in the lower two lines, the shape is concave.

However, plotting observed clear liquid heights actually yields a concave curve. Therefore, the Francis-type equations, with their convex curves, are inappropriate for expressing clear liquid height. It is safe to say that present equations for calculating clear liquid height are neither highly reliable nor highly realistic. Thus the need is great for an accurate and easy-to-use empirical method.

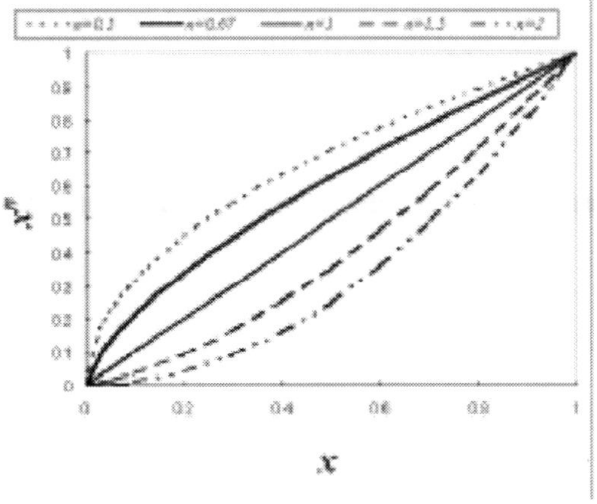

Fig. 21 The exponent 0.67 cannot express a concave curve

We modified the Bennett equation by changing its exponent of liquid load Q from 0.67 or 2/3 to about 1.5. Figures 22 and 23 compare the results obtained by the original and the modified Bennett equation. It is clear that the original Bennett equation cannot express the observed clear liquid height, whereas the modified Bennett equation can correlate the data fully.

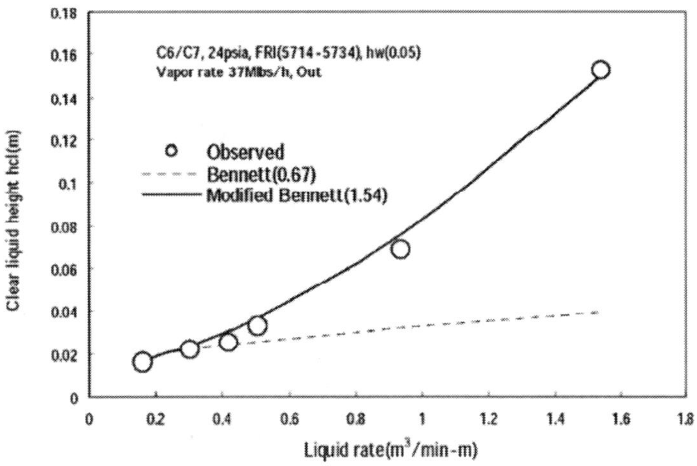

Fig. 22 The modified Bennett equation with exponent 1.54 can express a concave curve for clear liquid height

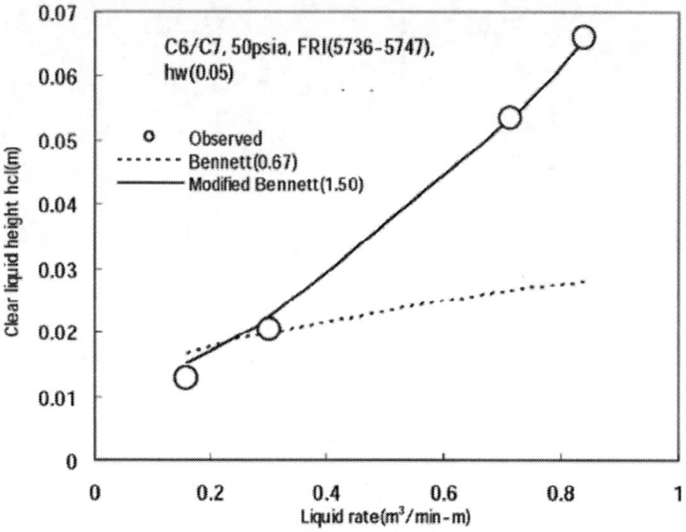

Fig. 23 The modified Bennett equation with exponent 1.50 can express a concave curve for clear liquid height

CONCLUSIONS

A new relation between the minimum entrainment and the newly defined residence time of liquid on the tray was found from an analysis of data observed at FRI. The liquid load at which the entrainment becomes minimum at vapor load constant operation coincides with the liquid load at which the newly defined residence time of liquid on the tray becomes minimum at the vapor load. In the newly defined residence time, volume is defined as the product of the bubbling area and the clear liquid height.

This concept affords a new, simple method for calculating the residence time. Its notable advantage is that it can predict the minimum entrainment value based solely on clear liquid height data. Presently available equations, such as the Francis type, do not yet adequately express clear liquid height. Development of more effective clear liquid height equations will make it possible to use this simple method, using residence time, to predict minimum entrainment. The modified Bennett equation by changing its exponent of liquid load Q from 0.67 or 2/3 to about 1.5 can express the observed clear liquid height.

ACKNOWLEDGMENTS

The author would like to express sincere appreciation to Mr. Tak Yanagi for his valuable collaboration in conducting the entrainment study. Thanks also go to graduate student T. Furuno and A. Nisiki for performing the clear liquid height calculations using the Bennett equation. Gratitude is expressed as well to FRI, for providing the experimental data.

NOMENCLATURE

C: defined by Eq. (2)
Cd : discharge coefficient
hcl : clear height, m
h_W : height of weir, m
Q_L : flow rate of liquid per unit length of weir, $m^3/s \cdot m$
V_B : superficial vapor velocity based on area of plate, m/s
W : length of weir, m
ε : gas holdup fraction
ε_W : value of ε in froth flowing over weir, εW=ε
α_e : effective liquid volume fraction
ρ_L , ρ^V : density, kg/m^3

REFERENCES

Bennett, D. L., R. Agrawal, and P. J. Cook, "New Pressure Drop Correlation for Sieve Tray Distillation Columns," AIChE J., 29, 434-442, 1983

Colwell, C. J., "Clear Liquid Height and Froth Density on Sieve Trays," Ind. Eng. Chem. Process Des. Dev., 20, 298-307, 1981

FRI, PROGRESS REPORT, MAY 1963

Kister, H. Z., "Distillation Design," McGraw-Hill, Inc. (1992).

Kister, H. Z, and J. R. Haas, "Entrainment from Sieve Trays in the Froth Regime," Ind. Eng. Chem. Res., 27, 2331 (1988).

Lockett, M. J., "Distillation Tray fundamentals, Cambridge University Press, Cambridge, U.K. (1986)

Pinczewski, W. V. and C. J. D. Fell, "The Transition from Froth-to-Spray Regime on Commercially Loaded Sieve Trays," Trans. Inst. Chem. Eng., 50, 102 (1972).

Pinczewski, W. V. and C. J. D. Fell, "Froth to Spray Transition on Sieve Trays," Ind. Eng. Chem. Process Des. Dev., 21, 774 (1982).

Sakata, M. and T. Yanagi, "Performance of a Commercial Scale Sieve Tray," Ind. Chem. Eng. Symp. Ser., 56, 3.2/21 (1979).

Sakata, M. and T. Yanagi, "Data Tables Supplement to Performance of A Commercial Scale Sieve Tray," Distillation 1979, I. Chem. E. Symposium Series No. 56 London, 3.2/21 (1979)

Silvey, F. C., T. R. Barben, M. Sakata, T. Yanagi, "Monthly Progress Report", FRI, May, 1963

Yanagi, T. and M. Sakata, "Data Tables Supplement to Performance of A Commercial Scale Fourteen Percent Hole Area Sieve Tray," presented for AIChE Spring National Meeting Distillation Tray Performance Symposium No. 44, Houston, Texas - April, 1981

129a Testing of Mellapakplus M452.Y at Fractionation Research, Inc

Mark W. Pilling[1], Florian Kehrer[2], and Lothar Spiegel[2]. (1) Sulzer Chemtech USA, Inc., 4019 S. Jackson Ave., Tulsa, OK 74107, (2) Sulzer Chemtech, Winterthur, Switzerland

MellapakPlus M452.Y high performance structured packing was test at Fractionation Research, Inc. (FRI) as part of their proprietary device testing program. This is the third type of MellapakPlus to be tested at FRI under this program. M252.Y and M752.Y were also tested.

The data from the M452.Y tests will be presented and compared with the previous data from M252.Y and M752.Y. The data should allow a complete analysis of the MellapakPlus family.

Discussion on the applications of the various types of MellapakPlus will be presented also.

129b Foaming Effect on Random Packing Performance

G.X. Chen[1], Tony Cai[1], **Karl T. Chuang**[2], and A Afacan[2]. (1) Fractionation Research, Inc., P.O. Box 2108, Stillwater, OK 74076, (2) Department of Chemical Engineering, University of Alberta, Edmonton, AB T6G 2G6, Canada

The effect of foaming on the performance of 16 mm Pall rings was determined in a 0.15 m distillation column with a bed depth of 1.23 m. The tests were conducted with the methanol/water system at 1.0 bar pressure. A controllable foaming inside the packed bed was generated successfully with the addition of the surfactant, Triton X-100, to the reflux stream. The performance of the packing with and without foaming was determined and compared. It was found that foaming increases the HETP by about 20 percent, reduces the capacity by approximately 30 percent. The results also show significant increase of bed pressure drop caused by foaming. In addition to the distillation tests, the foaming tendency and surface tension were measured for the methanol/water system at different methanol concentrations. Foaming tendency was measured using ASTM methods. The measurements revealed direct correlations between foaming tendency of the mixtures and the corresponding changes in surface tension.

Blended Packed Bed Performance SuperBlend 2-Pac™ Packing

Amistco Separation Products Inc.

Alvin, Texas

www.amistco.com

SuperBlend 2-Pac™ Packing

- **WHAT IS IT?**
- **Patented Invention**
  - » by Dr. B. J. Lerner
  - – A Blended Bed is a blended mixture of two sizes of AHPP packing or other random packings used as a single size bed would be used.

Amistco Separation Products Inc.

SuperBlend 2-Pac™ Packing

- **WHAT PERFORMANCE RESULTS?**
  - You get the best of both sizes
    - Bed Efficiency Equals The Smaller Size.
    - Bed Capacity & Pressure Drop of the Larger Size

SuperBlend-2Pac™ Packing

- **WHAT PROOF?**

- **Packing Tested by**
  - Separations Research Program (SRP) at The University of Texas – Dr. Frank Seibert
  - Distillation C6/C7 – Four Pressures
  - Air/H_2O Hydraulics
  - Absorption – CO_2/NaOH

Amistco Separation Products Inc.

Amistco High Performance Packing (AHPP™)

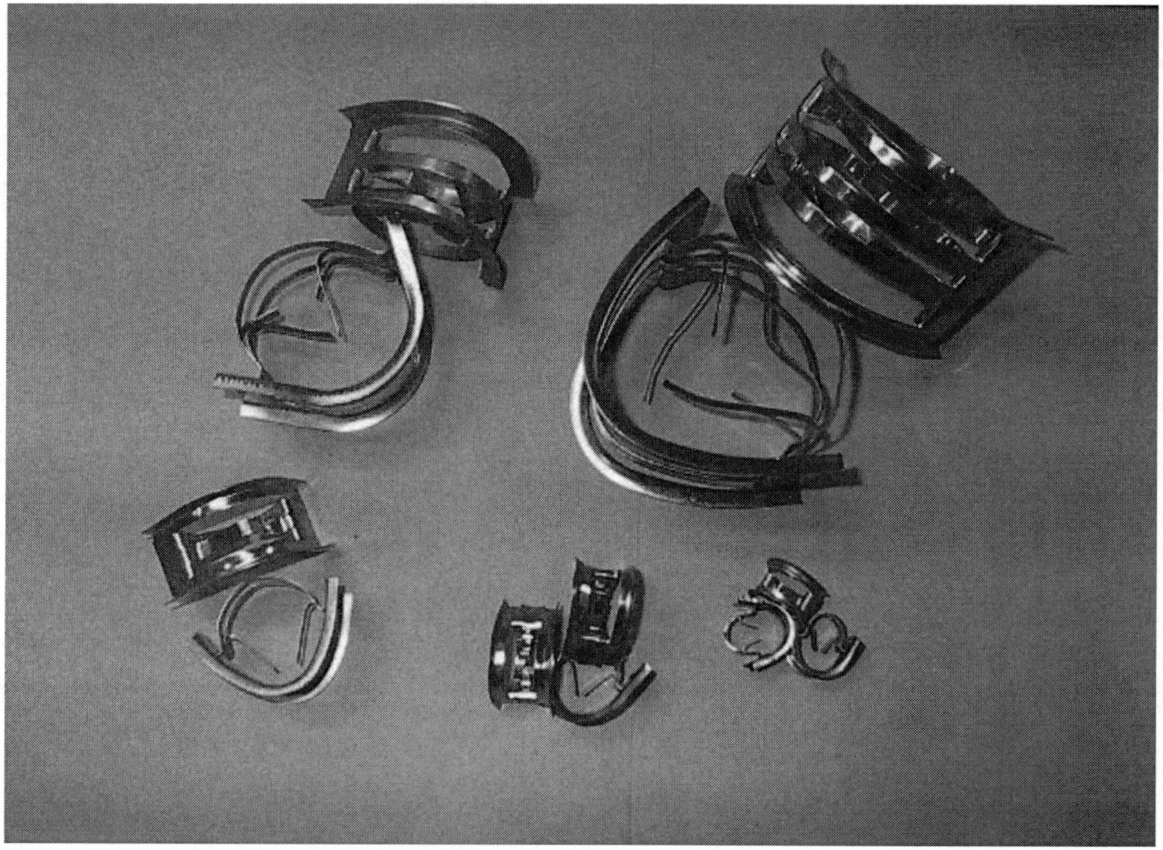

Amistco Separation Products Inc.

SuperBlend 2-Pac™ Packing

SB 2-Pac 2540 Packing Going Into SRP Tower

16.875" ID Tower by 10' Bed

Amistco Separation Products Inc.

SuperBlend 2-Pac™ Packing

Table 1

Distillation Data SB 2-Pac 2540 Pk			
System Psia	HETP, in Design	F_S Max. Avg. Top/Bot ft/sec(lb/ft^3)$^{0.5}$	$C_{S,}$Dip @ Tower Bot. Ft/sec
2.40	19-20	2.147	0.322
4.83	17.7	1.935	0.294
24	15.3	1.702	0.265
60	12.8	1.474	0.245

Amistco Separation Products Inc.

SuperBlend 2-Pac™ Packing

Table 2

Distillation Data SB 2-Pac 4050 Pk			
System Psia	HETP, in Design	F_s Max. Avg. Top/Bot $ft/sec(lb/ft^3)^{0.5}$	$C_{S,}$Dip @ Tower Bot. Ft/sec
2.40	25.3	2.75	0.409
4.83	23.9	2.579	0.391
24	20.7	2.204	0.350
60	17.2	1.8 Est.	-

Amistco Separation Products Inc.

SuperBlend 2-Pac™ 2540 Packing HETP vs Vapor Capacity

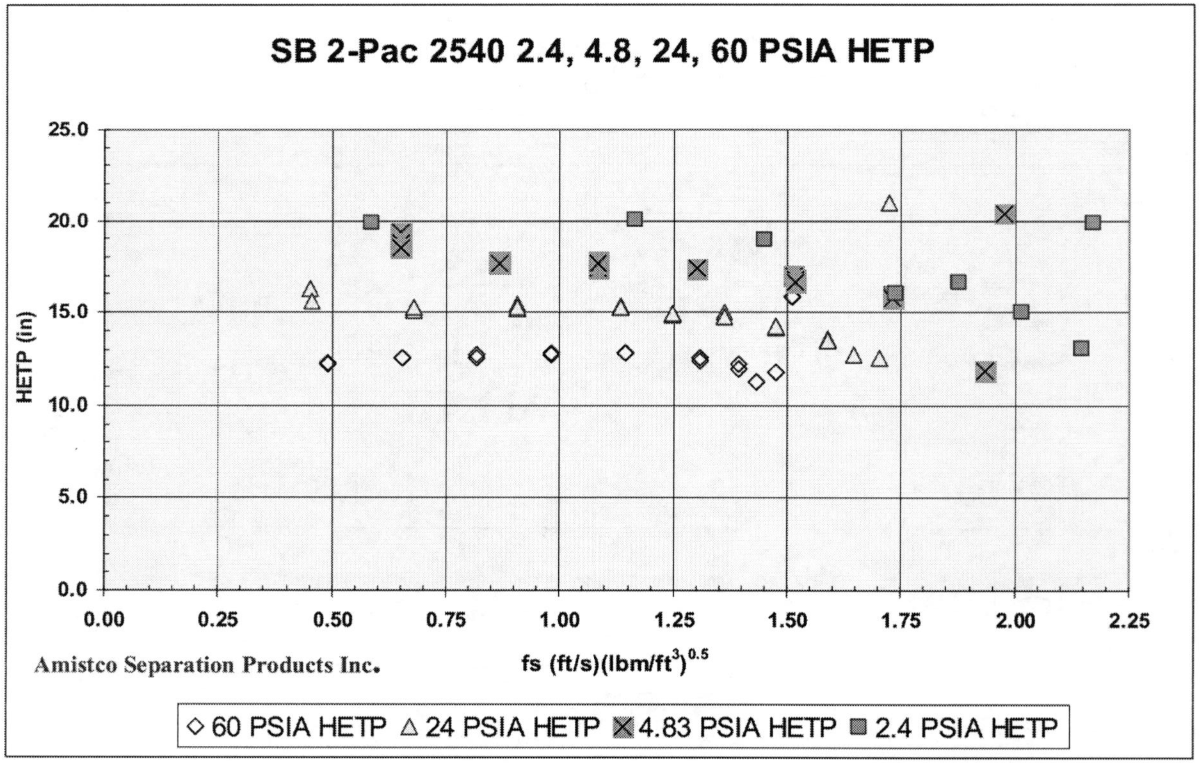

SuperBlend 2-Pac™ 4050 Packing HETP vs Vapor Capacity

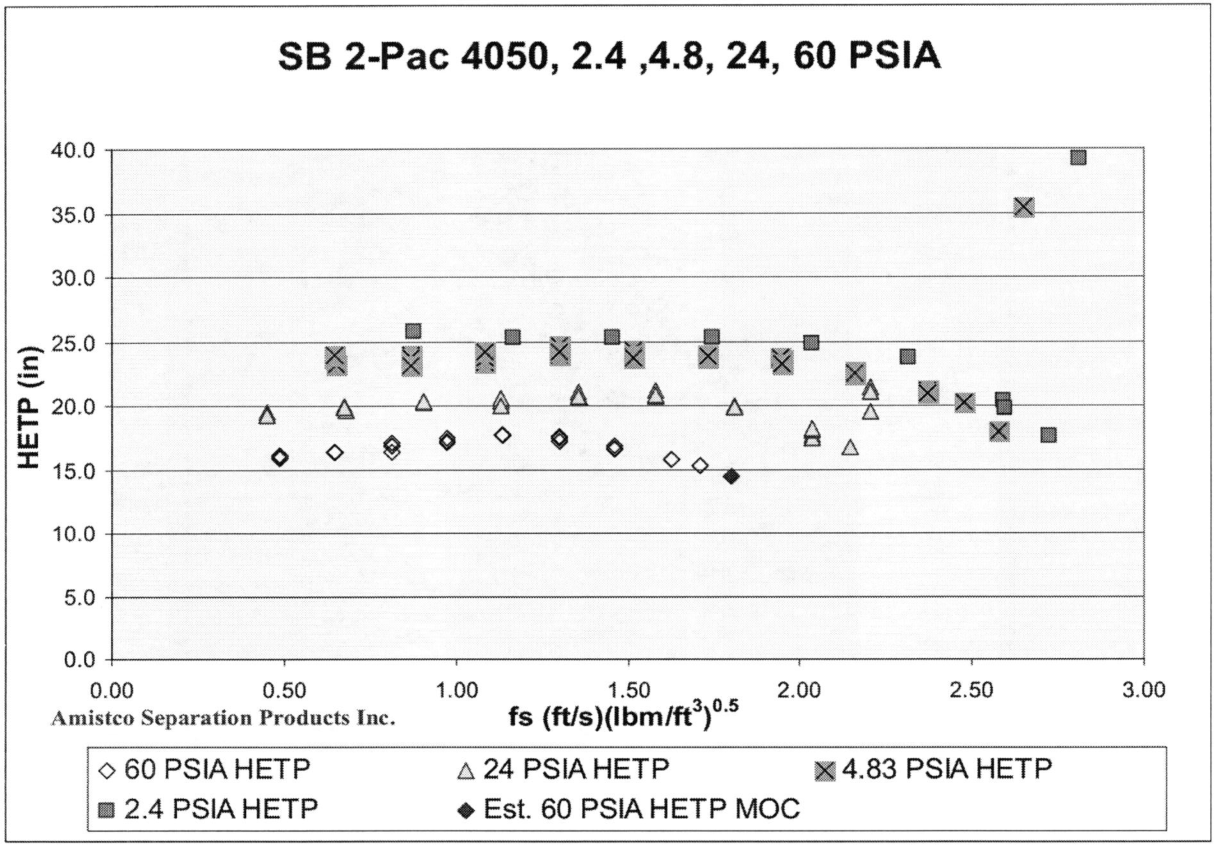

SuperBlend 2-Pac™ 2540 Packing

Figure 1, Air/H20 PD

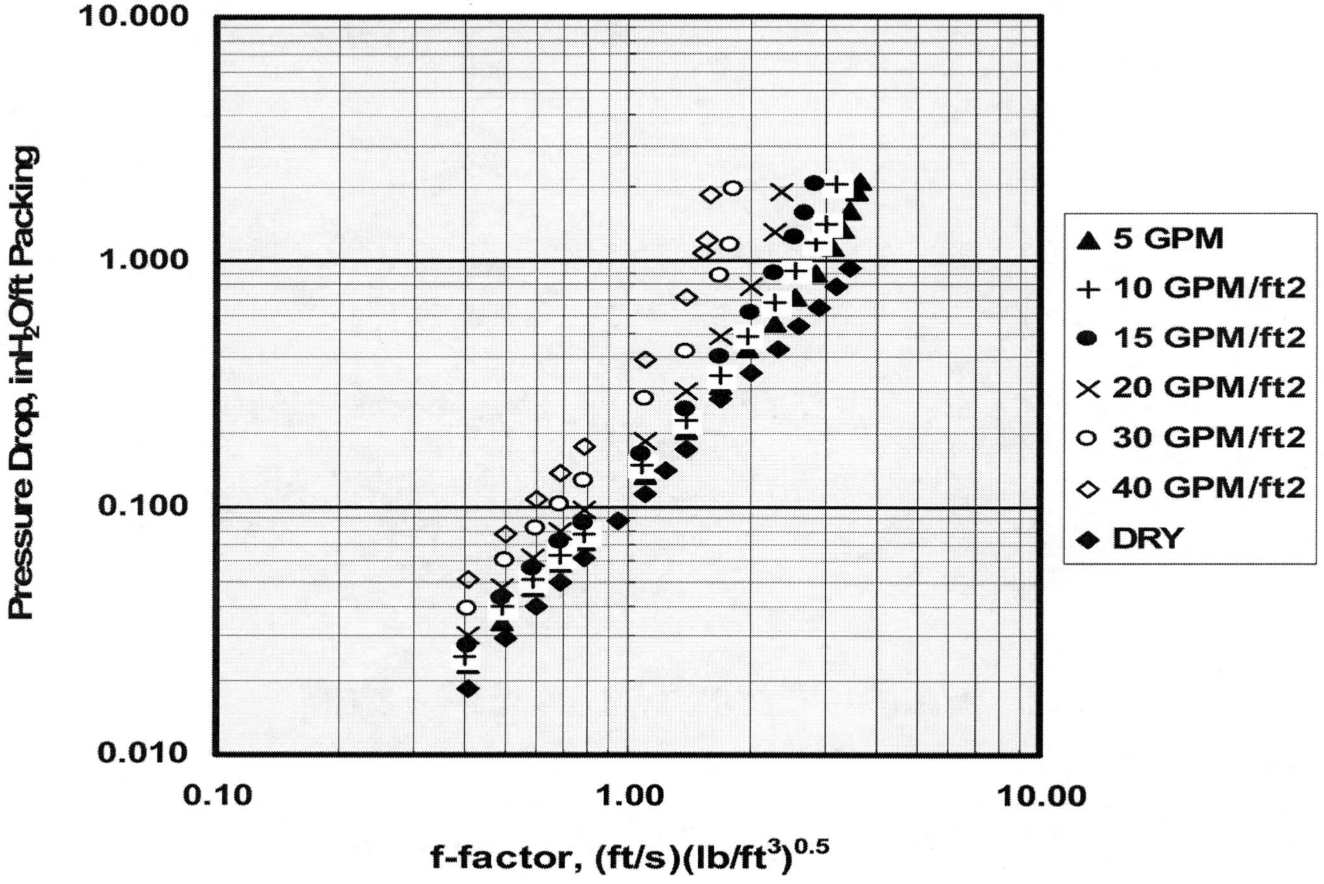

Amistco Separation Products Inc.

SuperBlend 2-Pac™ 4050 Packing

Figure 2, Air/H20 PD

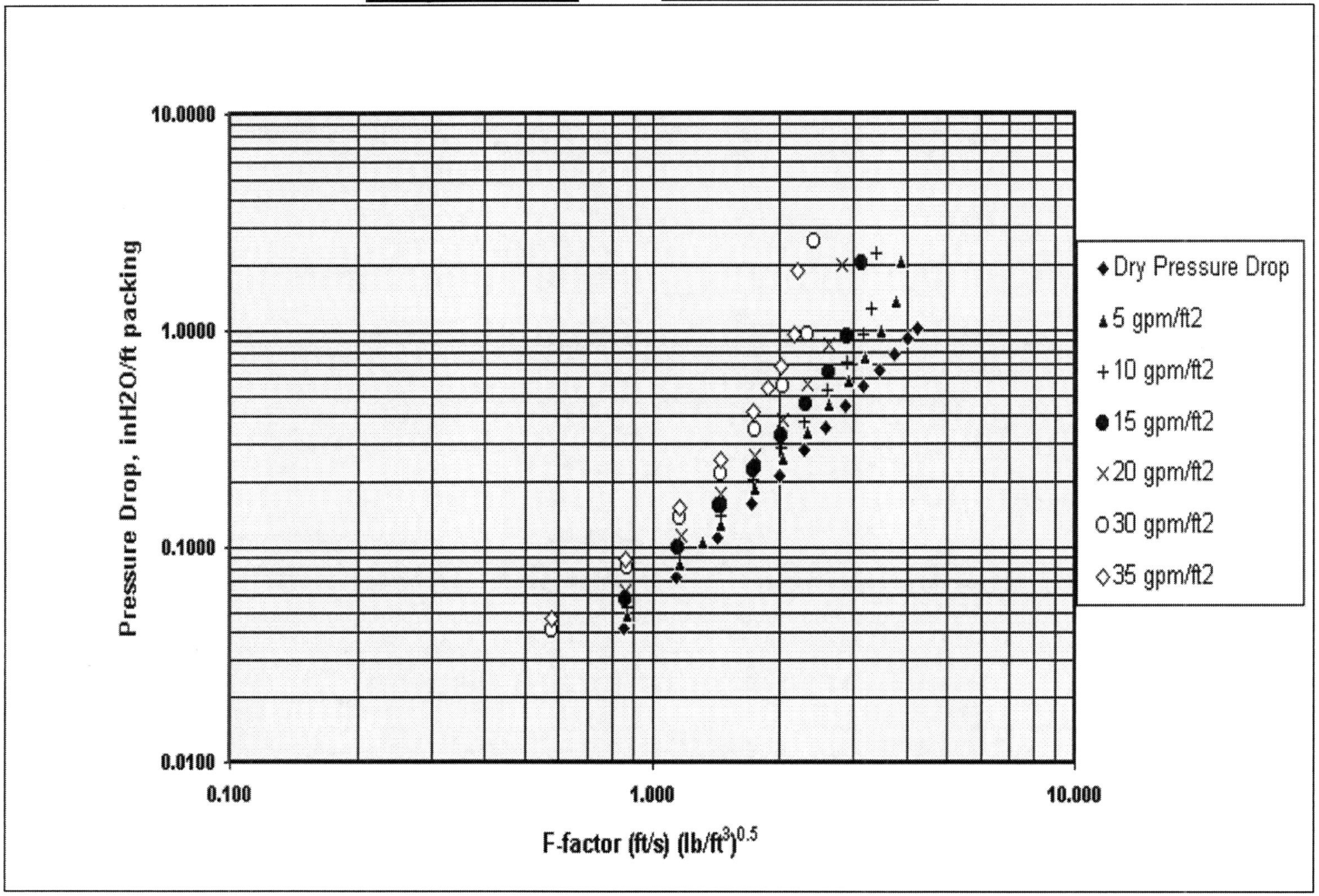

Amistco Separation Products Inc.

SuperBlend 2-Pac™ Packing Parity Plot Distillation Capacity

Figure 2, SB 2-Pac 2540 vs. IMTP™ 40

Cyclohexanen-Heptane - Total Reflux

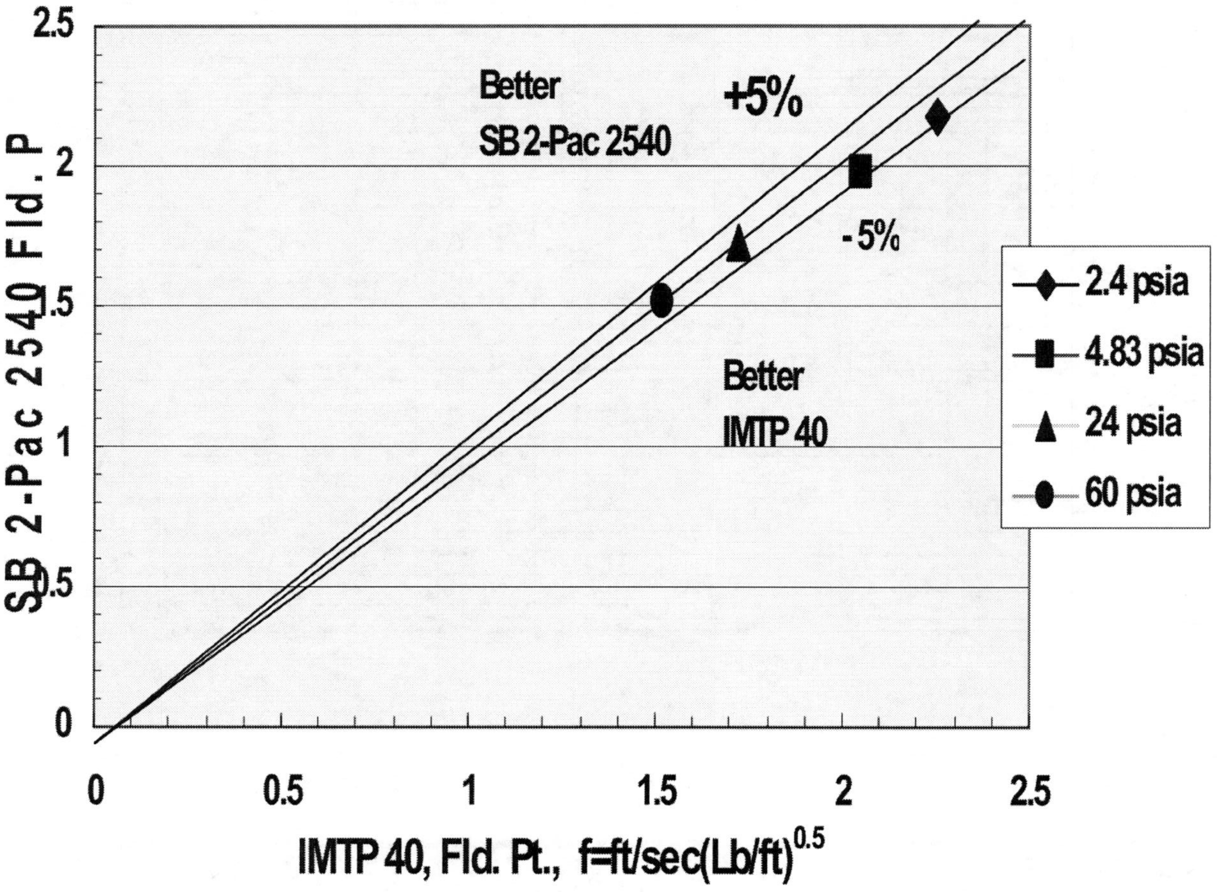

Amistco Separation Products Inc.

SuperBlend 2-Pac™ Packing Parity Plot Distillation Pressure Drop

Figure 3, SB 2-Pac 2540 vs. IMTP™ 40

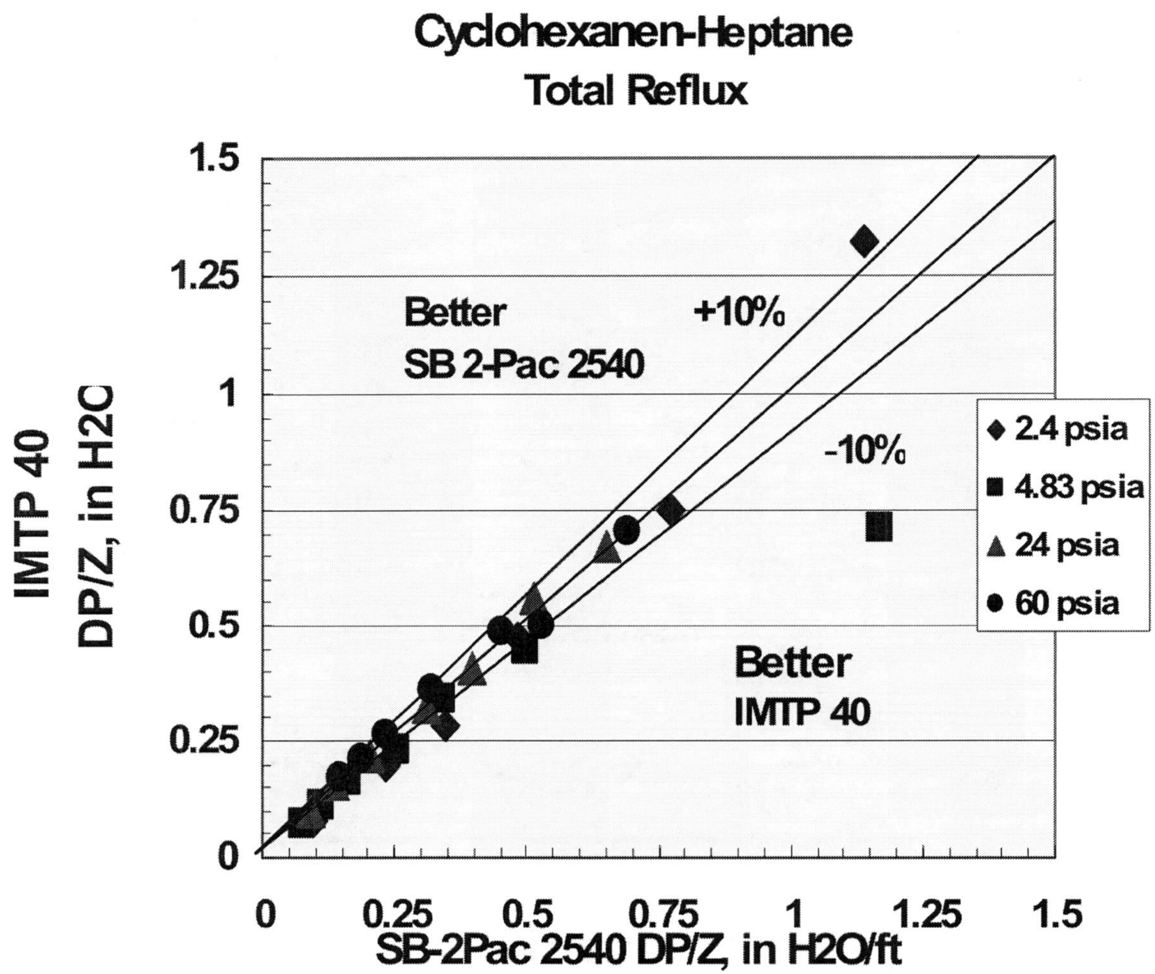

Amistco Separation Products Inc.

SuperBlend 2-Pac™ Packing Parity Plot Distillation HETP

Figure 4, SB 2-Pac 2540 vs. IMTP™ 40

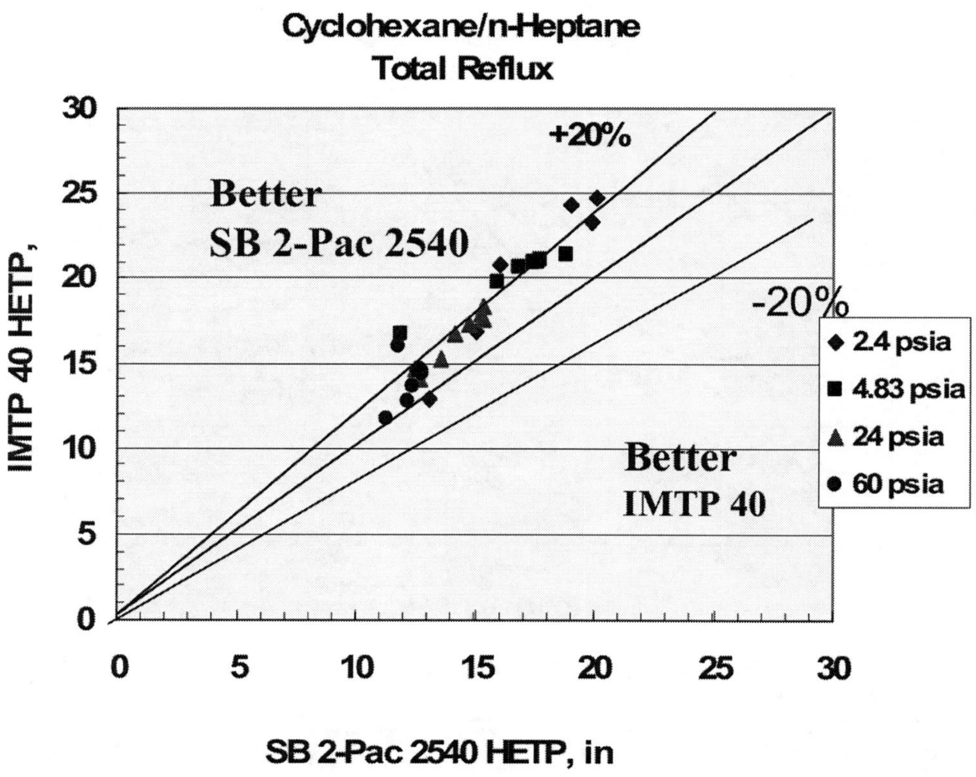

Amistco Separation Products Inc.

SuperBlend 2-Pac™ Packing Parity Plot Air-H_2O Pressure Drop

Figure 5, SB 2-Pac 2540 vs. IMTP™ 40

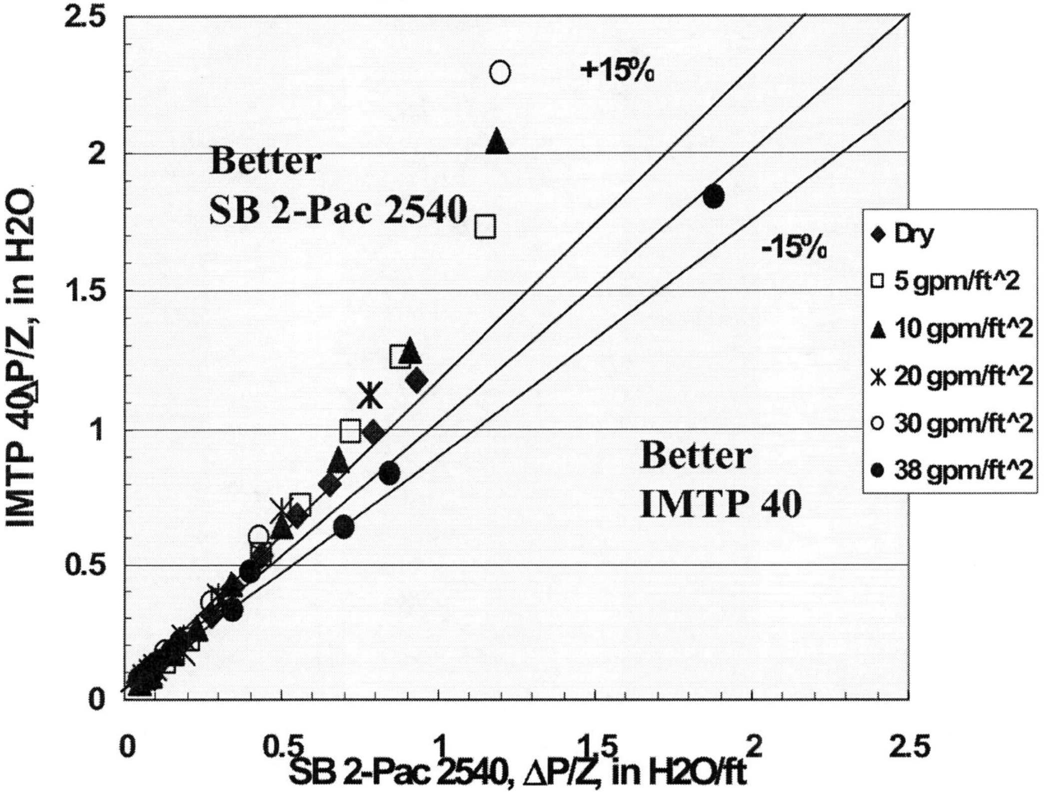

Amistco Separation Products Inc.

SuperBlend 2-Pac™ Packing Parity Plot Air-H_2O Flood Point

Figure 6, SB 2-Pac 2540 vs. IMTP™ 40

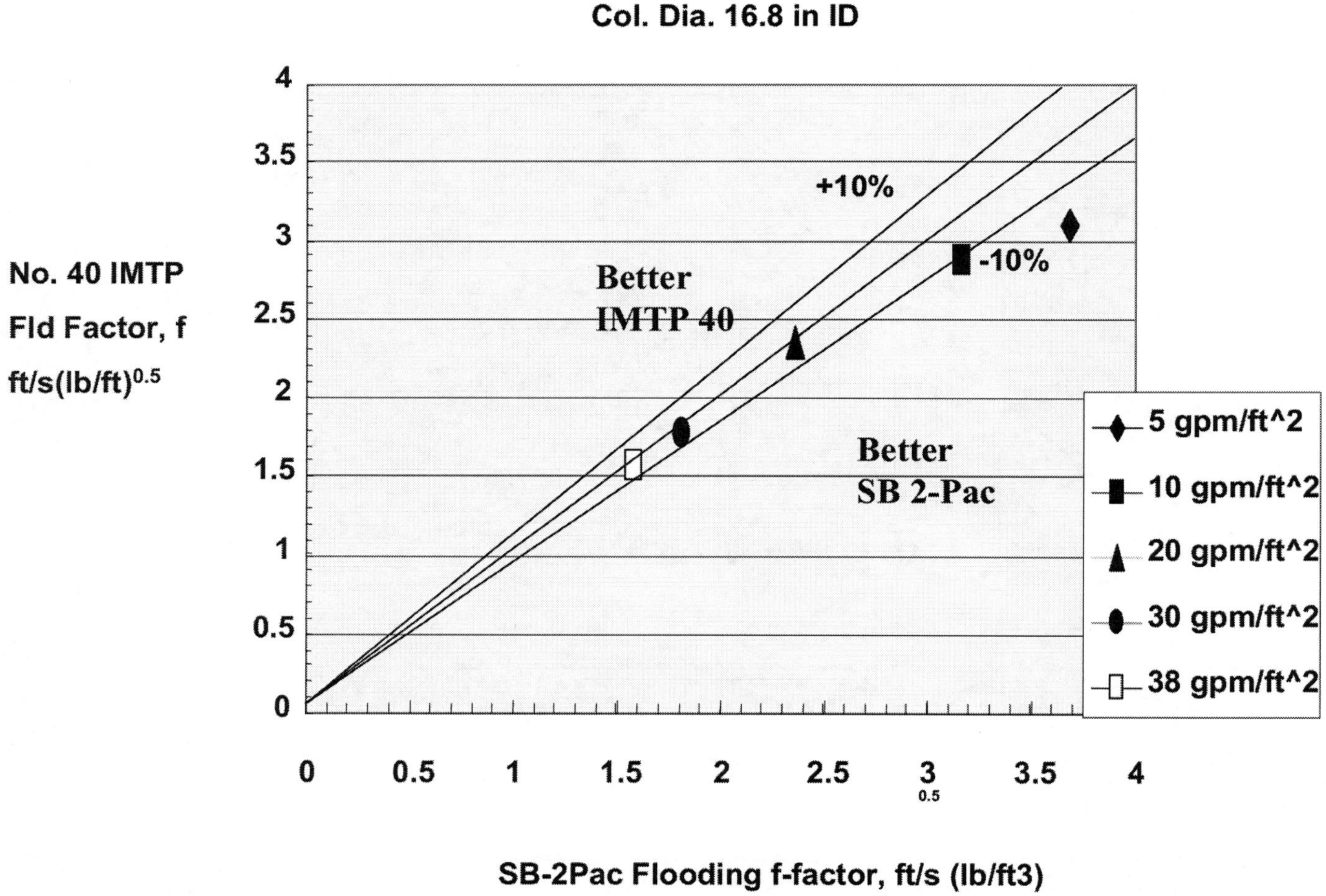

Amistco Separation Products Inc.

SuperBlend 2-Pac™ Packing Parity Plot Scrubbing Efficiency

Figure 7, SB 2-Pac 2540 vs. IMTP™ 25

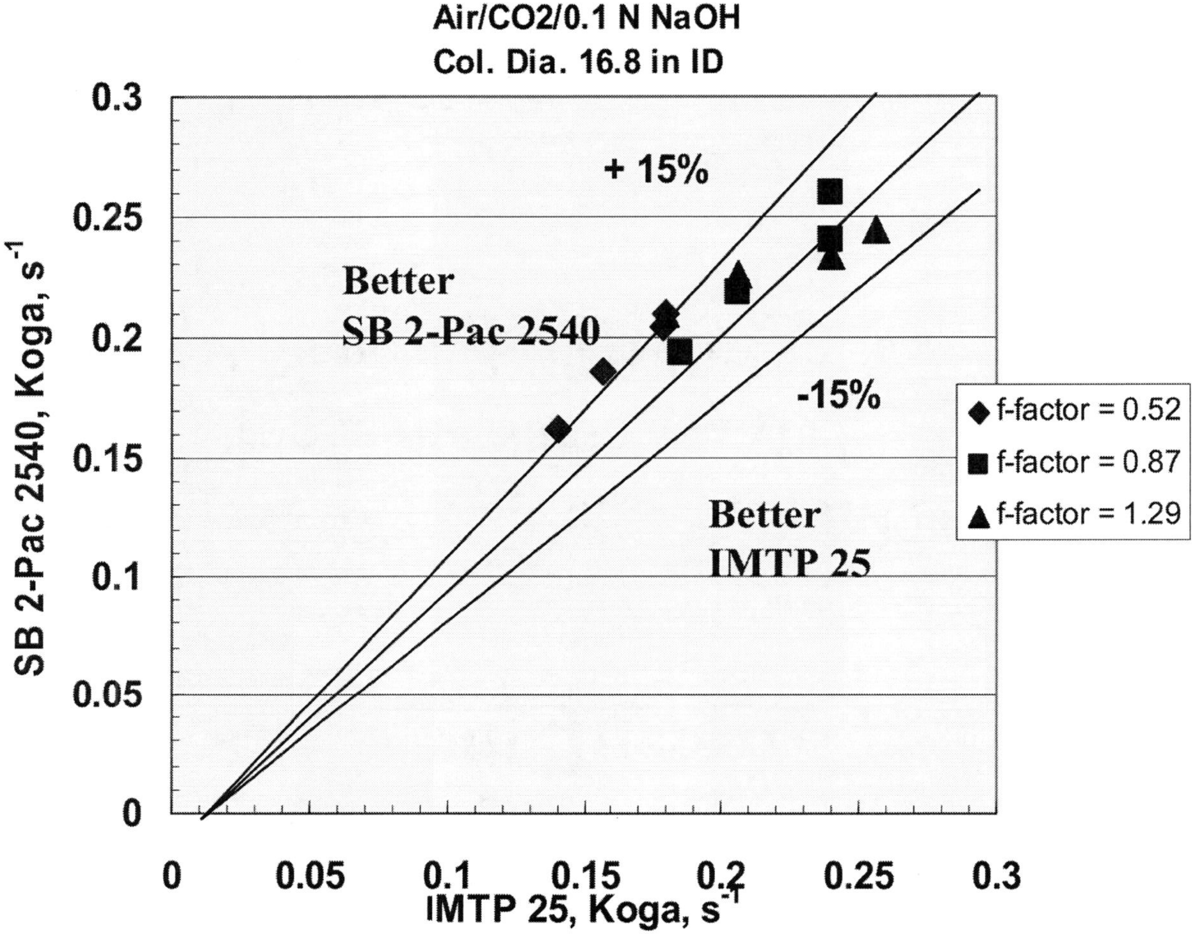

Amistco Separation Products Inc.

SuperBlend 2-Pac™ 2540 Packing Conclusions

- Distillation
  - HETP is 15% to 20% Better than IMTP™ 40, Fig. 4
  - Pressure Drop is the Same as IMTP 40, Fig. 3
  - Capacity is Same as IMTP 40, Fig. 2
- Absorption
  - Equal to IMTP 25, Fig. 7
- Air Water Hydraulics
  - Pressure Drop Equal to IMTP 40, Fig. 5
  - Flood Point Same as IMTP 40, Fig. 6

Amistco Separation Products Inc.

SuperBlend 2-Pac™ 2540 Packing Conclusions

- These are SRP's Summaries.
- Distillation Performance
  - "The distillation performance of the Amistco High Performance Random Packing (AHPP) achieved the equivalent Hydraulic Characteristics of the IMTP#40, a standard in high efficiency random packings, but with 25% greater distillation efficiency. A sharp increase in distillation efficiency was observed in the loading region."
- Absorption Performance
  - "The air/water hydraulic and mass transfer studies confirmed the findings of the distillation study. The Amistco High Performance Packing provides hydraulic characteristics similar to the IMTP#40 and mass transfer characteristics similar to the IMTP#25."

IMTP is a registered trademark of Koch-Glitsch, LLP

Amistco Separation Products Inc.

SuperBlend 2-Pac™ 4050 Packing Conclusions

- Detailed comparison to #40 IMTP™ still being made. First look indicates the HETP is about 5% greater than #40 IMTP™ and the capacity is significantly greater.
- No comparable data #50 IMTP™

IMTP is a registered trademark of Koch-Glitsch, LLP

Amistco Separation Products Inc.

SuperBlend 2-Pac™ 2540 Packing

C6/C7 - 60 PSIA

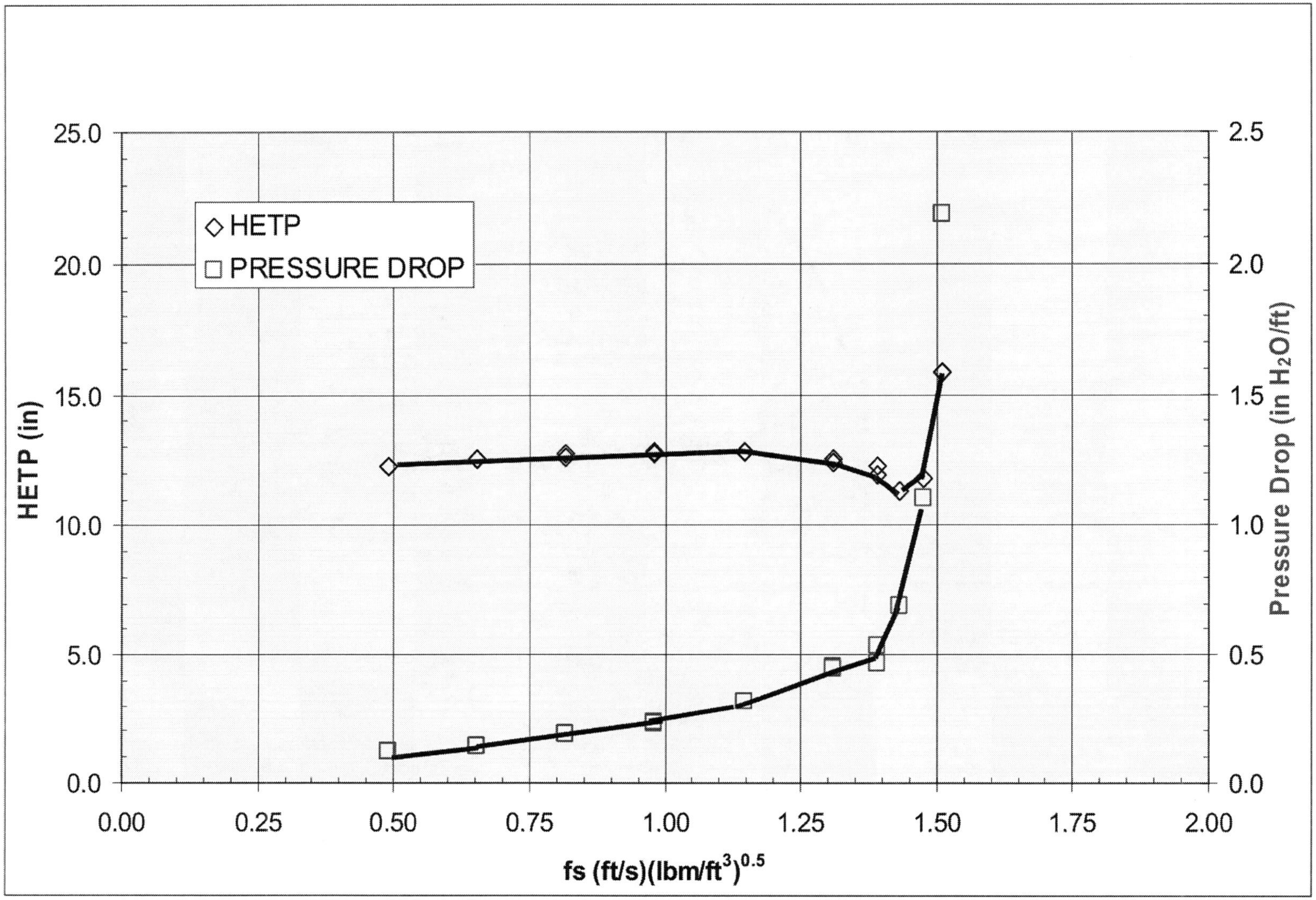

Amistco Separation Products Inc.

SuperBlend 2-Pac™ 2540 Packing

C6/C7-24 psia

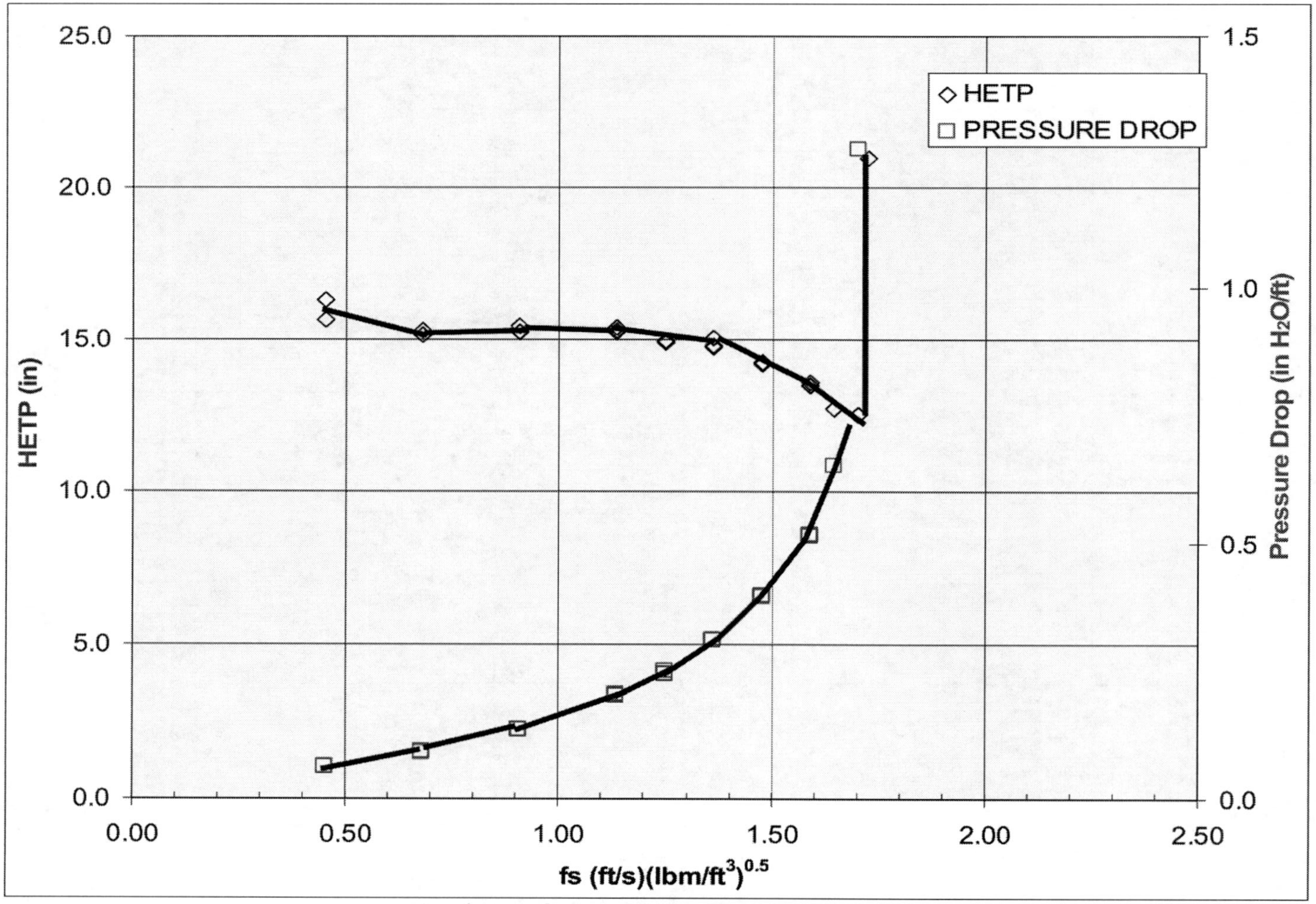

Amistco Separation Products Inc.

SuperBlend 2-Pac™ 2540 Packing

C6/C7-4.83 psia

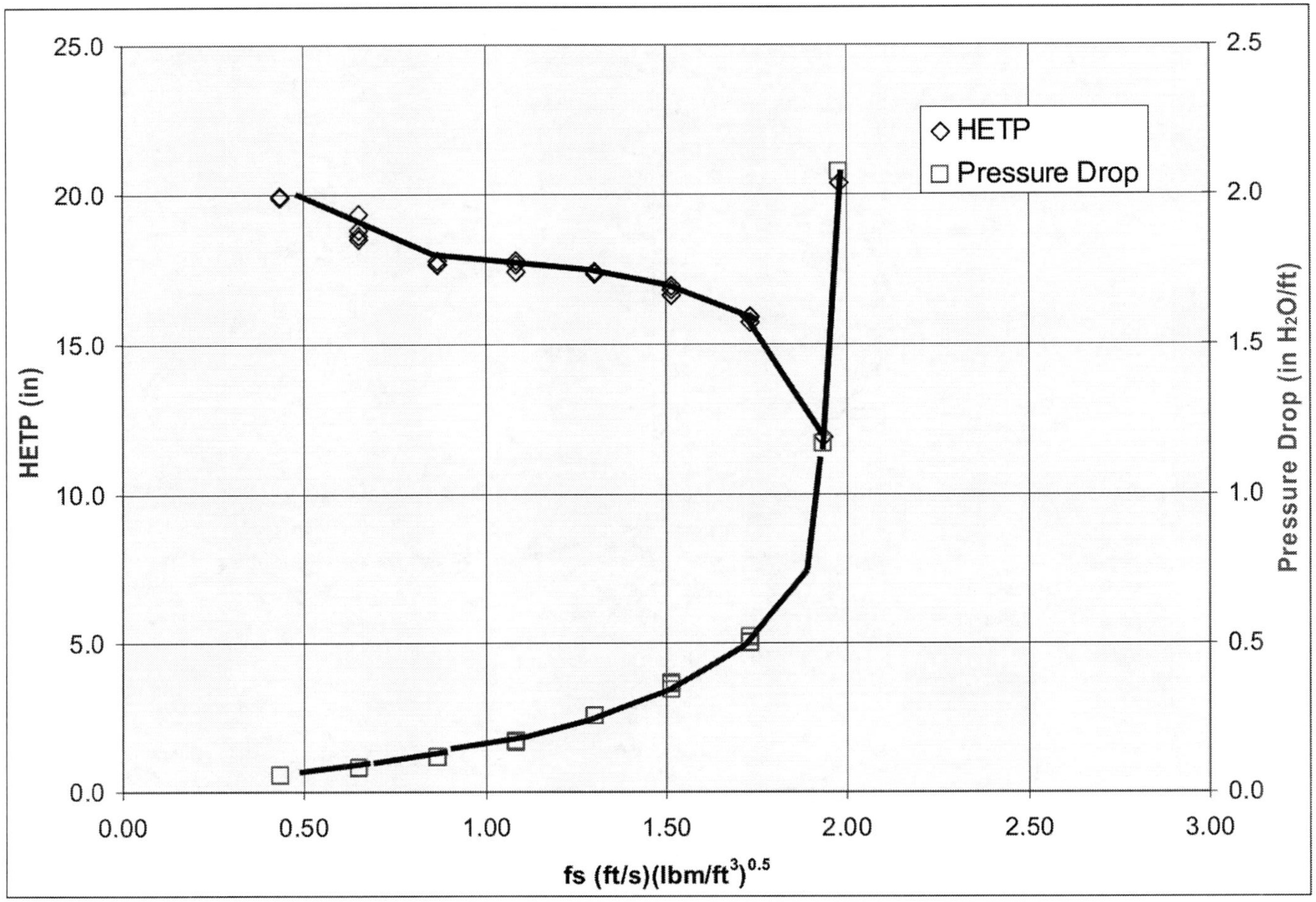

Amistco Separation Products Inc.

SuperBlend 2-Pac™ 2540 Packing

C6/C7- 2.40 psia

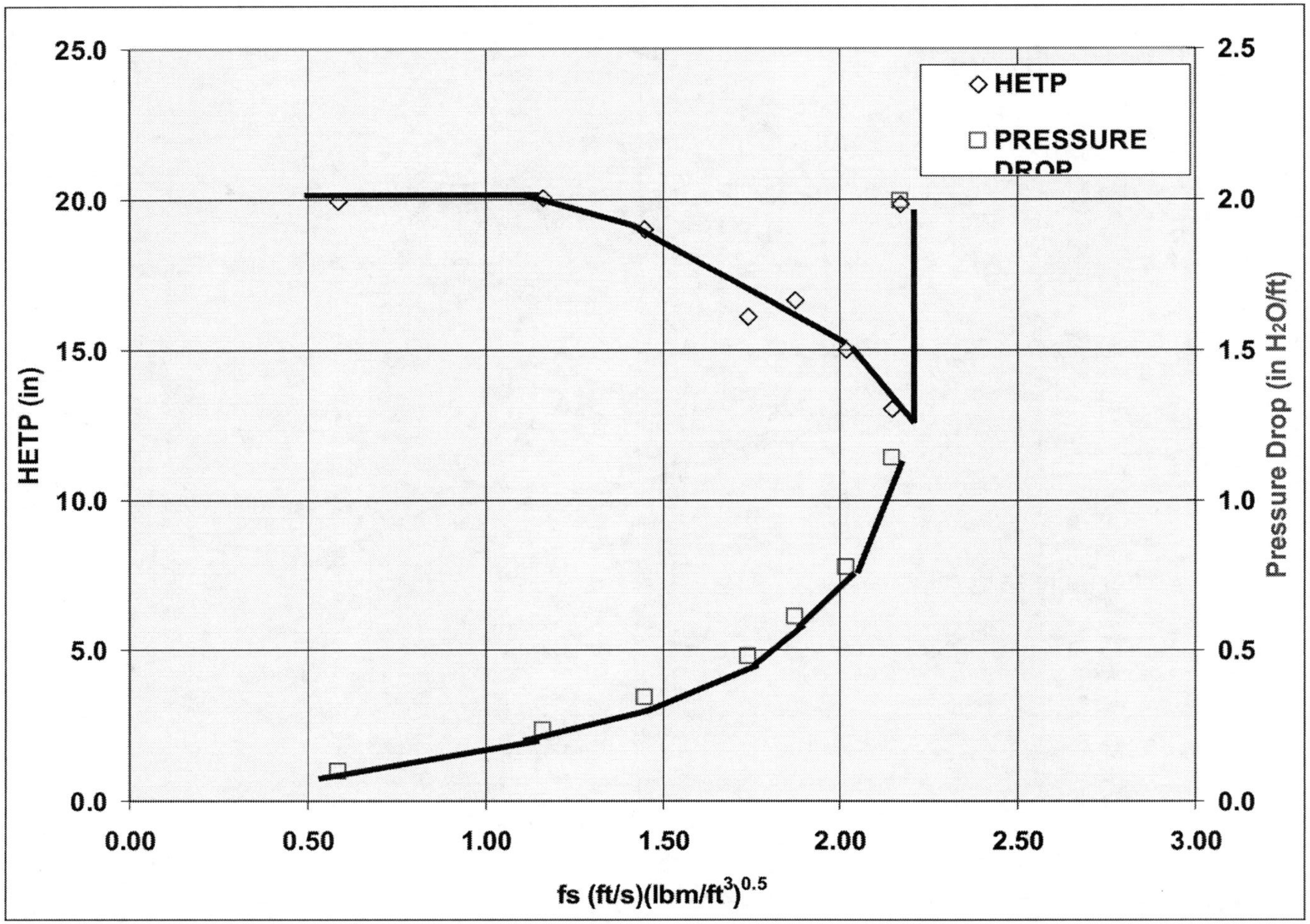

Amistco Separation Products Inc.

How to Surpass Conventional and High Capacity Structured Packings with Raschig Super-Pak

Michael Schultes, Simon Chambers
Raschig GmbH, Mundenheimerstrasse 100, 67061 Ludwigshafen, Germany.

Abstract

Raschig Super-Pak is a novel development in structured packing product technology that is fundamentally different to the well known corrugated perforated or non-perforated, textured sheet metal structured packings. It comprises of a systematic sequence of smooth sinusoidal waves above and below the plain of the metal sheet. Tests at the Ruhr University of Bochum showed that the new open structure resulted in very high capacities low pressure drops and excellent mass transfer efficiencies.

Encouraged by these early successes, total reflux distillation tests with Raschig Super-Pak 300 were conducted at the Separations Research Program, University of Texas at Austin. Results in terms of useful capacity, mass transfer efficiency, pressure drop and pressure drop per theoretical stage are presented. Comparisons with the Montz B1 250 and 250M and high capacity structured packings tested under identical conditions [1] will be shown. In addition results are compared against F.R.I. tested Mellapak and MellapakPlus structured packings for the same test fluid at 0.34 and 1.65 bar system pressures as reported by Pilling and Spiegel [2]. Overall Raschig Super-Pak 300 exhibits higher useful capacities and lower pressure drops compared not only with standard 45° corrugation angle $250m^2/m^3$ surface area structured packings but also with the high capacity counterpart types as well. Remarkably these gains were achieved without any adverse effects on mass transfer efficiency. This observation is reflected in the lower pressure drop per theoretical stage of Raschig Super-Pak 300 compared to the $250m^2/m^3$ surface area structured packings. These highly encouraging results provide industry with an alternative choice for smaller column diameter and/or higher capacity: the Raschig Super-Pak.

Keywords:
Raschig Super-Pak, Structured Packings, Distillation, Useful Capacity, Pressure Drop

1. Introduction

For the past 25 years structured packings have gained wide acceptance in the various chemical process industries, from deep vacuum (rectification) up to high pressure (absorption) because of their favourable capacity/low pressure drop/high efficiency characteristics. In response to higher capacity demands, modifications were required to standard structured packings because of initial flooding at the load point. It is well documented in the literature [3,4,5]. In brief at the load point, the liquid volume fraction is disproportionately higher at the packing layer interface where adjacent elements touch than in the bulk of packing elements. Combined with increased frictional forces imposed on rising vapour flow, the net result is significant restrictions on the liquid-vapour-traffic, which limit increases in capacity. Consequently, a new generation of high capacity structured packings, "Mellapakplus®", "Montz-PakM®" and "Flexipac®HC™" were developed. Specific details on how these new structured packing types were developed to increase capacity are explained elsewhere [3,4,5].

Despite the different performance characteristics between standard and high capacity structured packings, they all consist of vertically arranged bundles of corrugated (crimped), thin metal sheets. The most common corrugation angle to the horizontal is 45°, although a 60° angle is typically used when low pressure drop is of concern. Packing vendors use different surface enhancements on the thin metal sheets in an attempt to optimise wetting by use of holes, slots, vanes and tabs and/or lanced, grooved, textured or smooth treatment [6]. All methods attempt to improve turbulence within a liquid film, increase available surface area and promote mixing between different parts of the packing. The most common practice is to perforate individual metal sheets with the belief that it enhances uniform spread of vapour and/or liquid on the front and back of corrugated sheets within a packing layer. However Strigle [7] states that either liquid or vapour flow through perforations is a very low fraction of the total flow. Furthermore, it has been argued that perforating packing sheets can eliminate up to 10% of available material surface area and potential efficiency. This is supported by experimental evidence with a generic 250 m^2/m^3 structured packing in which the efficiency of the perforated version was about 8-10% lower than the non-perforated standard in the low liquid rate region [8].

2. Background

The purpose of this paper is to discuss development of Raschig Super-Pak 300 and how its performance compares with both conventional and high capacity corrugated sheet metal structured packings. Raschig Super-Pak 300 (RSP-300) is a novel development from Raschig GmbH comprising of a systematic sequence of smooth sinusoidal waves above and below the plain of the metal sheet that are arranged in rows at a 45° angle as shown in Figure 1.

Figure 1. Photograph of a Segment of Raschig GmbH Raschig Super-Pak 300 Structured Packing.

Production of the first Raschig Super-Pak yielded a surface area of 300 m^2/m^3 and is the first in a family of future planned Super-Pak structured packings. RSP-300 has a specific surface area of 301.7 m^2/m^3, void fraction of 0.977 and sheet metal thickness is 0.15 mm. Derived from hydraulic studies, the optimised open geometry deliberately promotes defined turbulent liquid film flows that spread uniformly between the back and front of each metal sheet via numerous contact points; a feature that is open to question with corrugated sheet types. The wave structure facilitates regular fluid communication between the back and front of each packing sheet, which maximises surface area and liquid/vapour distribution. This, in addition to lower shear stress forces encountered by countercurrent vapour flow, results in very high capacity, low pressure drop and excellent mass transfer performance.

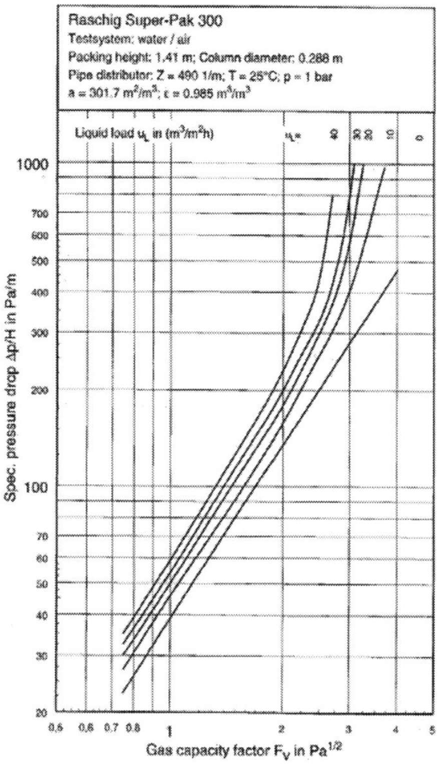

Figure 2. Effect of pressure drop per meter packed height of Raschig Super-Pak 300 with increasing gas capacity factor at constant liquid rates. Air-water tests Ruhr University, Bochum, Germany.

Extensive testing at the Ruhr University of Bochum with air-water and ammonia-air-water test systems show at least 25% higher capacity and 30% lower pressure drop compared with a conventional corrugated sheet metal structured packing with comparable specific surface area [9]. Results are shown in Figures 2 and 3.

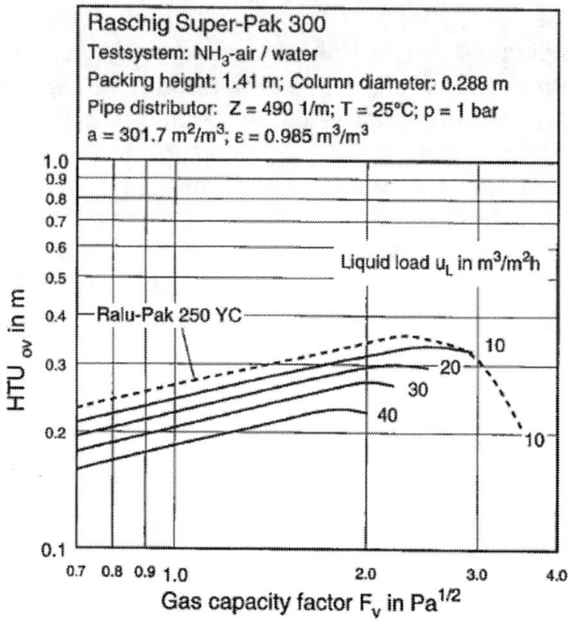

Figure 3. Effect of Height of a transfer unit of Raschig Super-Pak 300 with increasing gas capacity factor at constant liquid rates. Ammonia-air/water tests performed at Ruhr University, Bochum Germany.

3. Test Unit and Experimental Procedures

Encouraged by these results, the University of Texas at Austin Separations Research Program (SRP) conducted total reflux distillation tests to characterize the new RSP-300 metal structured packing. Hydraulic and mass transfer performance was measured using the cyclohexane/n-heptane (C_6/C_7) test system at operating pressures of 0.165, 0.33, 1.65 and 4.14 bar. Performance of the new RSP-300 structured packing is compared against the B1-250 and B1-250M conventional and high capacity structured packings from Montz tested under identical conditions. These data were taken from the Paper, "Performance of a New High Capacity Structured Packing" by Olujic et al [1]. In addition results are compared against F.R.I. tested Mellapak M250.Y and MellapakPlus M252.Y structured packings from Sulzer for the C_6/C_7 at 0.34 and 1.65 bar test systems as reported by Pilling and Spiegel [2].

Distillation tests were performed in the SRP 0.43 m ID column with a bed height of 3.124 m. The liquid distributor used was the SRP high capacity narrow trough drip tube distributor, with 145 pour points/m^2 and liquid flow rate range of 5 to 50 $m^3/m^2/h$. A complete description of the experimental set up and operating procedures can be found elsewhere [8].

Given the close-to-ideality nature of C_6/C_7 system at total reflux, the Fenske equation is used to calculate the number of equilibrium stages from the distillate and bottoms composition and average relative volatility. The average relative volatilities of the C_6/C_7 mixture together with various physical property values at the four operating pressures are given in Table 1.

It should be noted that the RSP-300 and Montz structured packing [1] experimental data were obtained using the SRP State-of-the-Art Fisher-Rosemount advanced process control and data acquisition system. This permits like-for-like comparisons.

Hydraulic results, in the form of pressure drop per unit height ($\Delta P/H$), pressure drop per theoretical stage (ΔP/Stage), and mass transfer results (HETP) are all plotted against vapour rate (F_S-Factor). F_S-Factor is an independent variable and is based on bottom of the column conditions.

Table 1. Physical Properties of the Cyclohexane/n-Heptane, C_6/C_7 System (Average at Column Bottom Conditions) at Four System Pressures Used in Total Reflux Tests

Pressure (bar)	**0.165**	**0.33**	**1.65**	**4.14**
Liquid Density, kg/m3	659	656	608	560
Liquid Viscosity, mPa-s	4.5E-1	4.3E-1	2.3E-1	1.6E-1
Liquid diffusivity, m2/s	2.29E-9	2.73E-9	6.19E-9	9.16E-9
Vapor Density, kg/m3	0.65	1.18	5.44	13.1
Vapor viscosity, mPa-s	6.6E-3	7.0E-3	8.3E-3	9.1E-3
Vapor diffusivity, m2/s	13.4E-6	11.3E-6	2.94E-6	1.42E-6
Surface tension, mN/m	18	17	14	8
Relative volatility	1.96	1.80	1.57	1.42
Average Temperature, °C	49	61	114	152

4. Results and Discussion

4.1 Liquid Rate and Operating Pressure Effect on Efficiency

Figure 4 shows good and stable HETP values for RSP-300 at all four operating pressures over almost the entire liquid rate range of the high capacity liquid distributor. With each operating pressure, HETP reaches a constant value in the preloading regime. In the loading regime there is a pronounced decline in HETP values, indicative of improved mass transfer efficiency, prior to a sharp break in the HETP curve as the packing enters incipient flood. The exception was at 4.14 bar because the reboiler capacity had reached its upper limit prior to flooding. It can be seen the HETP is generally between 0.305 and 0.410 m regardless of pressure.

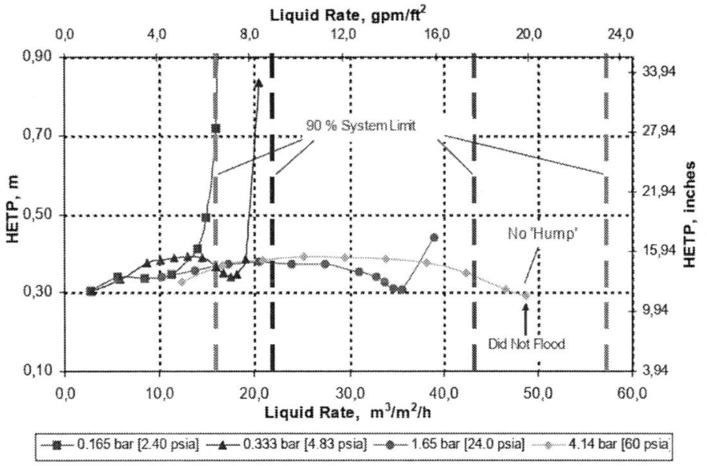

Figure 4. Effect of liquid rate and operating pressure on Raschig Super-Pak 300 HETP at total reflux. 0.43 m I.D. SRP Column, 3.124 m Bed, C6/C7 System, High capacity distributor with drip tubes

At 4.14 bar system pressure and liquid rate of 48.7 $m^3/m^2/h$, typical of high pressure distillation, there was no "efficiency hump", a phenomenon observed with both conventional and high capacity structured packings [1,10). The result is remarkable given that the flow channels in adjacent RSP-300 packing layers are at 90° with respect to one another when vertically stacked. The open structure helps alleviate any restrictions in vapour-liquid flows and possible vapour backmixing that otherwise may be found in the more closed channels of corrugated sheet structured packings.

Figure 4 has superimposed on it the 90% System Limit for each of the four operating pressures and will set a precedent for the remaining figures below. At higher pressures, the rise in RSP-300 HETP at flow rates close to hydraulic flood occurs long before the 90% System Limit is reached. This is clearly shown with 1.65 bar pressure. With 4.14 bar pressure, the rise in RSP-300 HETP at hydraulic flood would have occurred long before the 90% System Limit had the reboiler not reached its capacity limit. With decreasing pressure, the rise in the RSP-300 HETP curve at flow rates close to hydraulic flood converge and rapidly approach the 90% System Limit until a critical point is reached where it crosses the 90% limit. This is illustrated with 0.165 bar pressure in Figure 4. Overall, it implies that as lower the operating pressure, the increased tendency of high performance structured packings such as Raschig Super-Pak to rapidly approach 90% System Limit before entering full hydraulic film flood.

4.2 Efficiency and Useful Capacity

Capacity-efficiency comparative plots of RSP-300 with B1-250 and B1-250M structured packings, taken from the Paper "Performance of a New High Capacity Structured Packing"[1], are presented in Figures 5 and 6 with the C_6/C_7 at 1.65 and 0.33 bar test systems. RSP-300 hydraulic and mass transfer data at 1.65 bar operating pressure are compared against B1-250 and B1-250M packing measurements at 1.03 bar since no runs were made at 1.65 bar. In Figure 5 the HETP in the mid capacity range at 1.65 bar pressure for RSP-300 is 0.375 m compared to 0.36 m and 0.39 m for the B1-250 and B1-250M respectively. At 0.33 bar, Figure 6 shows an HETP of 0.38 m for RSP-300 in the mid capacity range compared to 0.36 m for B1-250 and 0.41 m for B1-250M.

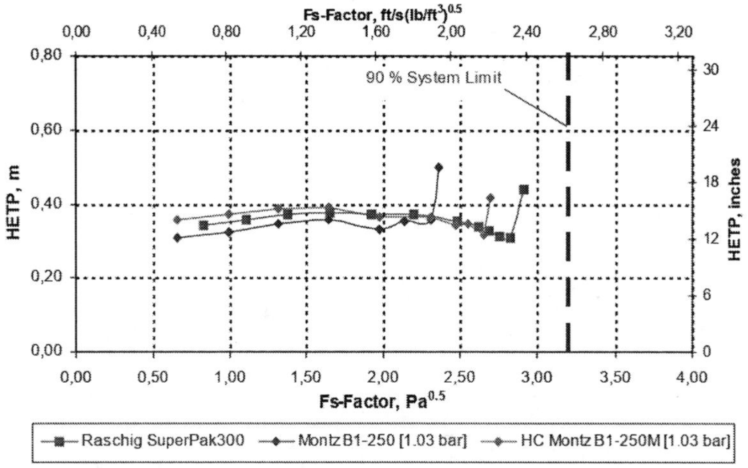

Figure 5. Mass transfer efficiency (HETP) comparison at total reflux. Raschig Super-Pak300 vs B1-250 and B1-250M at 1.65 bar, 0.43 m I.D. SRP column, 3.124 m bed, C_6/C_7 system, High capacity distributor with drip tubes

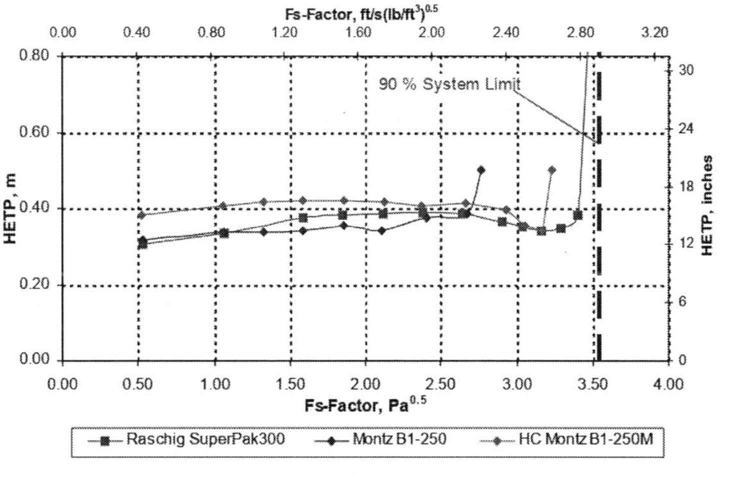

Figure 6. Mass transfer efficiency (HETP) comparison at total reflux. Raschig Super-Pak300 vs B1-250 and B1-250M at 0.33 bar, 0.43 m I.D. SRP column, 3.124 m bed, C_6/C_7 system, High capacity distributor with drip tubes

For both operating pressures, RSP-300 shows both a distinct minimum in HETP, and a clear maximum useful capacity advantage compared with B1-250. At 1.65 bar, Figure 5 shows a maximum useful capacity advantage of 22% for RSP-300 compared with B1-250 and more significantly 7% over the high capacity B1-250M. Had B1-250 and B1-250M been tested at 1.65 bar the useful capacity advantages may have been greater. Similarly at 0.33 bar in Figure 6, the RSP-300 useful capacity advantage is 37% compared to B1-250 and 7% over B1-250M.

Figures 7 and 8 compare HETP of RSP-300 with F.R.I. tested Mellapak M250.Y[10,2] and high capacity MellapakPlus M252.Y[2].

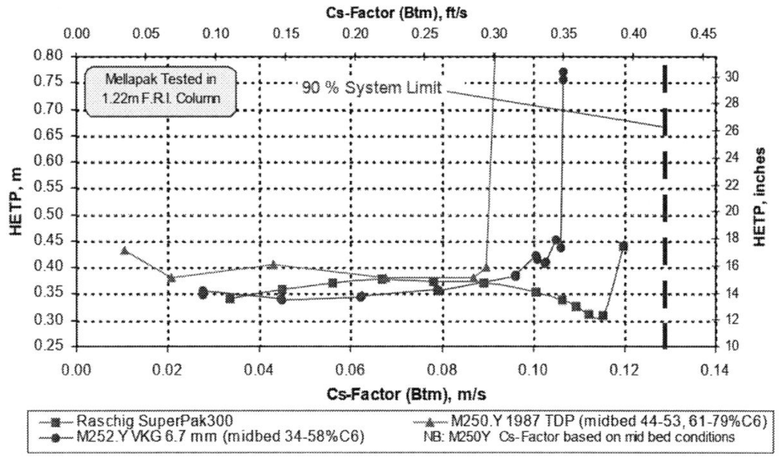

Figure 7. Mass transfer efficiency (HETP) comparison at total reflux. Raschig Super-Pak300 vs.F.R.I. tested M250Y and M252Y at 1.65 bar, 0.43 m I.D. SRP column, 1.22m F.R.I. column, C_6/C_7 system, High capacity liquid distributors

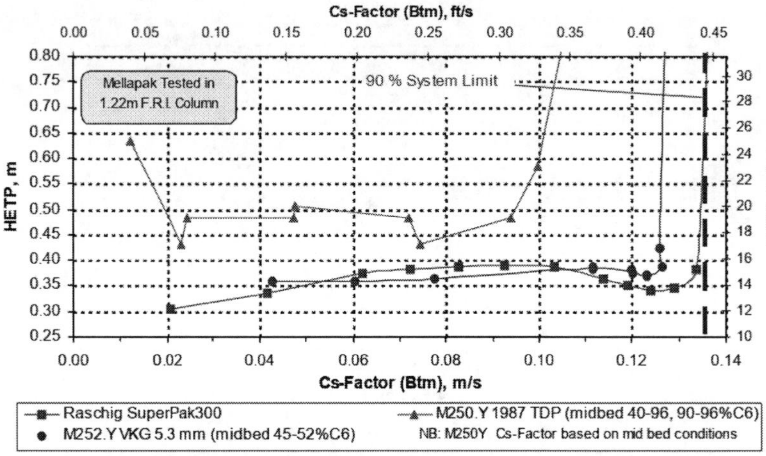

Figure 8. Mass transfer efficiency (HETP) comparison at total reflux. Raschig Super-Pak300 vs.F.R.I. tested M250Y and M252Y at 0.33 bar, 0.43 m I.D. SRP column, 1.22m F.R.I. column, C_6/C_7 system, High capacity liquid distributors

Both plots utilize Cs-Factors based on column bottom conditions and mid-bed C6 composition range except for M250.Y since no data were available in the original tests to calculate Cs from bottom column conditions. With this in mind, the HETP in the mid capacity range for M250.Y and M252.Y at 1.65 bar (Figure 7) are 0.39 m and 0.35 m respectively compared to the RSP-300 HETP of 0.375 m. At 0.33 bar, the HETP in the mid capacity range for M250.Y and M252.Y are 0.48 m and 0.37 m respectively compared with the RSP-300 HETP of 0.38 m as shown in Figure 8. As with the B1–250 packing, RSP-300 displays a substantial maximum useful capacity advantage compared with M250.Y. This is illustrated in Figures 7 and 8 with maximum useful capacity advantages of 28% and 41% for RSP-300 compared with M250.Y at the respective operating pressures of 1.65 and 0.33 bar. When compared against the high capacity M252.Y, the useful capacity gains of RSP-300 at 1.65 and 0.33 bar are 9% and 5% respectively; a significant result. All HETPs and useful capacities are summarised in Table 2 where B1-250 is taken as a reference point set at 100% for a given liquid and vapour rate, against which all the other packings are compared.

Table 2. Performance Comparison Raschig Super-Pak 300 versus B1-250, M250.Y,standard and B1-250M and M252.Y High Capacity Structured Packings at Total Reflux Conditions with C_6/C_7 System

Pressure p [bar]	Valuation	Montz B1-250	Sulzer M250 Y	Montz B1-250M	Sulzer M252 Y	Raschig RSP 300
1.65 (1.03)*	$F_{S,Flood}$ [$Pa^{1/2}$]	100%	106%	115%	113%	123%
	$F_{S,max}$ [$Pa^{1/2}$] Δp = 5 mbar/m	100%	101%	121%	128%	136%
	$F_{S,max}$ [$Pa^{1/2}$] Δp = 3 mbar/m	100%	104%	118%	129%	139%
	Mid Capacity Range HETP [m]	0.36	0.38	0.39	0.35	0.375
0.33	$F_{S,Flood}$ [$Pa^{1/2}$]	100%	102%	113%	121%	124%
	$F_{S,max}$ [$Pa^{1/2}$] Δp = 5 mbar/m	100%	119%	128%	138%	145%
	$F_{S,max}$ [$Pa^{1/2}$] Δp = 3 mbar/m	100%	121%	129%	146%	142%
	Mid Capacity Range HETP [m]	0.36	0.48	0.41	0.37	0.375

Notes:
** Extracted from data reported by Pilling and Spiegel [2]
* Montz B1-250 and B1-250M tested at 1.03 bar [1] not 1.65 bar.
With conventional structured packings Useful Capacity corresponds to standard design ΔP of 3 mbar/m.
With high capacity structured packings, the shift in Useful Capacity towards higher pressure drop corresponds to arbitrary design ΔP of 5 mbar/m.

The useful capacity advantage of RSP-300 over the high capacity B1-250M and M252.Y structured packings is remarkable given that the flow channels in adjacent RSP-300 packing layers are at 90° with respect to one another when vertically stacked. The open structure helps alleviate any restrictions in vapour-liquid flows at the element interface where layers touch.

4.3 Hydraulic-Pressure Drop Comparison

Pressure drop comparative plots of RSP-300 with B1-250 and B1-250M packings are presented in Figure 9 for 1.65 and 0.33 bar operating pressures. For both system pressures RSP-300 pressure drop is considerably lower than both the B1-250 and B1-250M structured packing over the entire operating range. To quantify the useful capacity and pressure drop advantages of Raschig Super-Pak 300 over B1-250, we will use 3 mbar/m as a reference point since it is typically the design value for standard structured packings. With high capacity structured packings, there is a shift in useful capacity to higher pressure drop values (3 up to 5 or 6 mbar/m). As a result we will compare useful capacity change of RSP-300 with the B1-250M at 5.0 mbar/m reference point. Figure 9 indicates 43% and 13% useful capacity advantages for RSP-300 over B1-250 and B1-250M at 1.65 bar. At 0.33 bar pressure, RSP-300 displays useful capacity gains of 43% and 11% compared with B1-250 and B1-250M. At the lower operating pressure, the ΔP curve at maximum useful capacity for RSP-300 is very close to or at the 90% System Limit. Had B1-250 and B1-250M been tested at 1.65 bar the useful capacity advantages may have been greater.

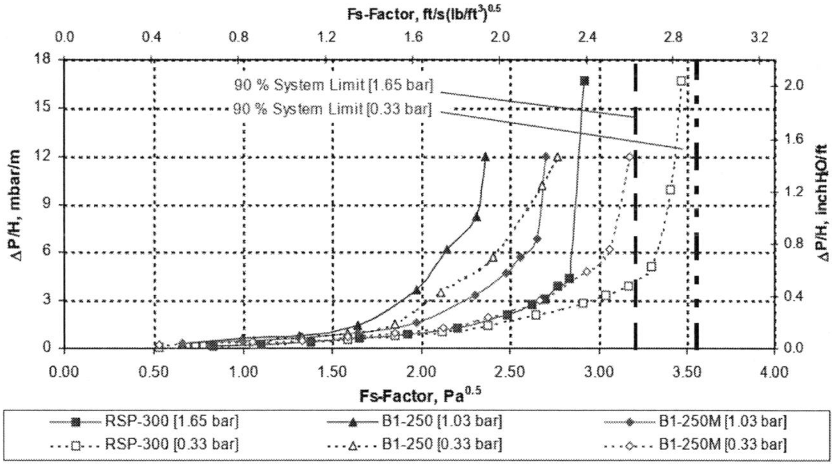

Figure 9. Pressure Drop comparison at total reflux. Raschig Super-Pak300 vs.B1-250 and B1-250M at 1.65 and 0.33 bar, 0.43 m I.D. SRP column, 3.124 m bed, C_6/C_7 system,
High capacity distributor with drip tubes

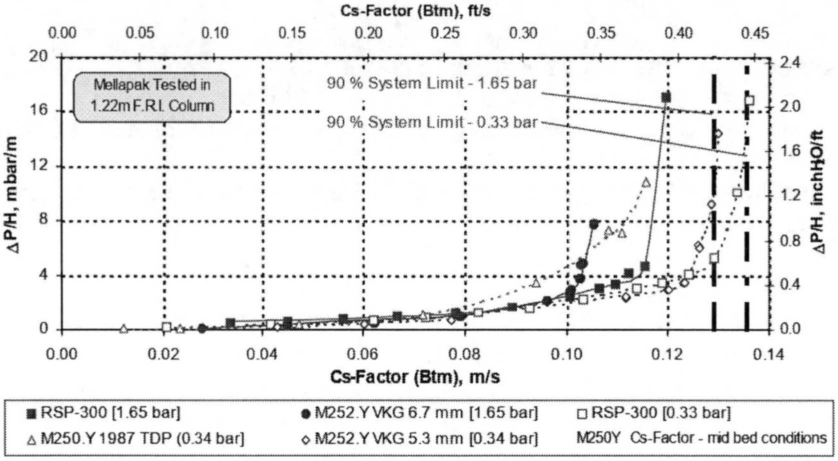

Figure 10. Pressure drop (ΔP/H) comparison at total reflux. Raschig Super-Pak300 vs.F.R.I. tested M250Y and M252Y at 1.65 and 0.33 bar, 0.43 m I.D. SRP column, 1.22m F.R.I. column, C_6/C_7 system, High capacity liquid distributors

Figure 10 compares the pressure drop of RSP-300 with M250Y and M252Y at 1.65 to 0.33 bar operating pressures. As with the HETP plots, Cs-Factors based on column bottom conditions and mid-bed C6 composition range have been utilized except for M250.Y for reasons given above. Additionally no pressure drop data were taken at 1.65 bar for M250.Y test in 1987. On comparing useful capacity at the design reference point of 3 mbar/m for the standard M250.Y and 5 mbar/m for the high capacity M252.Y, RSP-300 at 0.33 bar has 41% more useful capacity than M250Y and 5% more than M252.Y. The convergence of maximum useful capacity of both RSP-300 and M252Y at the lower pressure, suggests that the on-set of hydraulic film flooding is very close to or at the 90% System Limit for both packings. At 1.65 bar RSP-300 has a more pronounced useful capacity advantage of 9% over the high capacity M252Y; a remarkable result. Like the B1 packings, all HETPs and useful capacities for M250.Y and M252.Y are summarised in Table 2.

With structured packing applications there is on the whole a trade-off between capacity and efficiency. Consequently, pressure drop per theoretical stage is an important parameter in evaluating different structured packing designs. Figure 11 shows the pressure drop per theoretical stage comparisons of RSP-300 with B1-250 and 250 standard and high capacity structured packings at 1.65 and 0.33 bar operating pressures. Likewise Figure 12 shows the pressure drop per theoretical stage comparisons of RSP-300 with M250.Y and M252.Y. For both pressures, the pressure drops per theoretical stage of RSP-300 is consistently and distinctly lower than the B1-250 and B1-250M in Figure 11. In Figure 12, RSP-300 shows a noticeably lower pressure drop per theoretical stage over M250Y and M252Y in the high capacity operating range for both operating pressures. These results are very favourable in low pressure and vacuum columns processing thermally sensitive fluids.

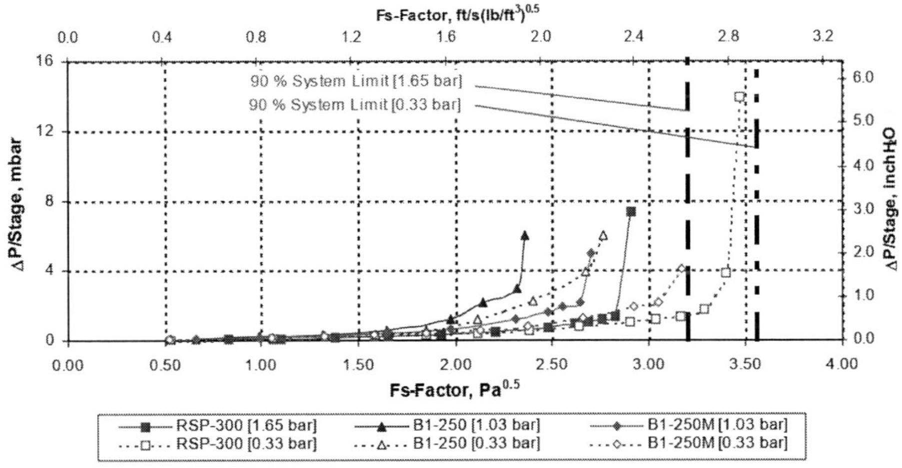

Figure 11. Pressure Drop per Theoretical Stage comparison at total reflux. Raschig Super-Pak300 vs.B1-250 and B1-250M at 1.65 and 0.33 bar, 0.43 m I.D. SRP column, 3.124 m bed, C_6/C_7 system,

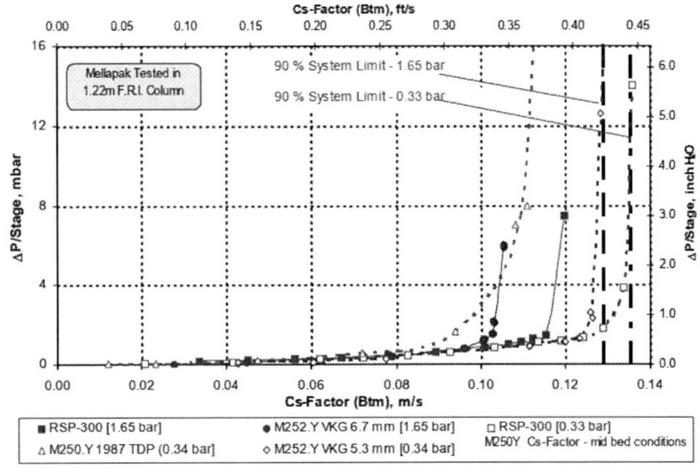

Figure 12. Pressure Drop per Theoretical Stage comparison at total reflux. Raschig Super-Pak300 vs.F.R.I. tested M250Y and M252Y at 1.65 and 0.33 bar, 0.43 m I.D. SRP column, 1.22m F.R.I. column, C_6/C_7 system

On the whole, the excellent hydraulic advantages of RSP-300 over the B1-250M and M252.Y high capacity structured packings is remarkable given that adjacent Raschig Super-Pak 300 layers are rotated 90° with respect to one another. This is in contrast to the B-250M and M252.Y high capacity types with curvature of the corrugations gradually changing from 45° to 0° on vertical axis at one or both ends of the packing elements.

5. Conclusions

The novel design of Raschig Super-Pak 300 structured packing offers the various chemical process industries an alternate choice over the well known corrugated sheet metal conventional and high capacity structured packings. The open structure results in excellent hydraulic and mass transfer efficiency characteristics as verified in the total reflux distillation tests at the Separations Research Program (SRP), University of Texas at Austin.

Significant useful capacity and low pressure drop advantages were obtained not only over the standard B1-250 and M250.Y structured packings but over the B1-250M and M252Y high capacity structured packings as well. Equally important is that mass transfer efficiency was at least as good as if not better than the 250m^2/m^3 surface area structured packings. At low operating pressures, the convergence of hydraulic performance of the high capacity structured packings with that of RSP-300, suggests that the onset of hydraulic flood is very close to or at 90% System Limit. This was not observed at higher operating pressures of 1.65 bar and above. The very encouraging hydraulic advantages of RSP-300 over the B1-250M and M252Y high capacity structured packings is remarkable given that adjacent Raschig Super-Pak 300 layers are rotated 90° with respect to one another like standard structured packings. This contrasts with the gradually changing corrugation angle from 45° to 0° in the vertical axis, which characterises the B1-250M and M252Y high capacity structured packings. The implications for RSP-300 are that in new column design the required diameter will decrease with increasing the capacity of the packing used in a column providing the mass transfer efficiency remains constant. Alternatively debottlenecking an existing column to boost throughput capacity can be achieved by replacing the existing packing with higher capacity packing: the Raschig Super-Pak.

Acknowledgements

The authors would like to acknowledge the efforts of Prof. Fair, Dr. Frank Seibert , Mr. Christopher Lewis and the SRP staff for their excellent work on the Raschig Super-Pak 300 testing program.

6. References

[1] Olujic, Z., Seibert, A.F., Kaibel, B., Jansen, H., Rietfort, T., Zich. E., "Performance of a New High Capacity Structured Packing", AIChE Spring National Meeting, Houston, TX, 22-26 April, 2001.

[2] Pilling, M., Spiegel, L., "Design Characteristics and Test Validation of High Performance Structured Packing", Distillation Symposium, Distillation Honors John Kunesh, AIChE November National Meeting, Reno Nevada, November 2001.

[3] Spiegel, L., Meier, W., "Structured Packings Today and Tomorrow", Proceedings of Vision 2020: Distillation Priorities Topical Symposium, AIChE Spring National Meeting, Atlanta, March 5-9, 2000.

[4] Olujic, Z,, Jansen, H., Kaibel, B., Rietfort, T., Zich. E., "Stretching the Capacity of Structured Packing"; Ind. Eng. Chem. Res. No. 40 (2001), pp 6172-6180.

[5] McNulty, K. J., Sommerfeldt, R. A., "New Twists Adds Capacity to FLEXIPAC Structured Packing", Topical Conference Preprints, Distillation: Horizons for the New Millennium, AIChE Spring National Meeting, Houston, TX, March 14-18, 1999.

[6] Kister, H. Z.: Distillation Design, McGraw-Hill Inc. 1992.

[7] Strigle, R. F., Jr., Packed Tower Design and Applications: Random and Structured Packings, 2nd Edition, Gulf Publishing, Houston, TX, 1994, pp 41.

[8] Olujic, Z., Seibert, A. F., Kaibel, B., Jansen, H., Rietfort, T., Zich. E., "Effect of Perforations on the Performance of High Capacity Structured Packings with Large Surface Area", AIChE Spring National Meeting, New Orleans, 30 March - 2 April, 2003.

[9] Raschig Super-Pak 300: A New Packing Structure with Innovative Advantages, Bulletin No. VRR-5200-20d/IV/98/So, 1998.

[10] Fitz, C. W., Kunesh, J. G., and Shariat, A., "Performance Of Structured Packing in a Commercial-Scale Column At Pressures of 0.02-27.6 Bar", I.& E.C. Research, 38, 1999, p. 512.

Nomenclature

C_6		Cyclohexane
C_7		n-Heptane
C_S	m/s	Gas or vapour capacity factor = $F_S / \sqrt{\rho_L - \rho_V}$
F_S	$\sqrt{Pa}$ or $m/s(kg/m^3)^{1/2}$	Gas or vapour capacity factor = $u_S \cdot \sqrt{\rho_V}$
H	m	Packing height
HETP	m	height equivalent to a theoretical stage
L	kg/h	Liquid mass flow rate
p	bar	operating / system pressure
u_L	m^3/m^2h	superficial liquid velocity
u_S	m/s	superficial vapour velocity
V	kg/h	Gas or vapour mass flow rate

Greek letters

ΔP	Pa, mbar	Pressure drop
ρ_L	kg/m^3	Liquid density
ρ_V	kg/m^3	Gas or vapour density

129e Radioisotope Technology - Benefits & Limitations in Packed Bed Tower Diagnostics

Scott Vidrine, Tracerco, Suite 200, 1100 Hercules, Houston, TX 77058

The use of radioisotope technology in Chemical, Petrochemical and Refinery Process Plant mass transfer vessels has continuously been developed over the past 50 years to assist engineers in troubleshooting process problems, optimize production and plan shutdown schedules to minimize plant down time. Many advances in data acquisition and processing ability now allow critical information to be provided quickly to ensure the most efficient use of plant equipment.

With the advancement of data gathering and handling it is important to note that fundamental physics behind each and every measurement involving radioisotopes has never changed. As with any technology there are many advantages in its use. However there are also limitations that a process engineer must fully understand so that a possible problem is not overlooked or simply discounted based upon results generated.

Mass transfer design can typically be split into two main categories, trayed and packed beds. Packed beds are becoming increasingly popular. In the case of a trayed tower, the open area is typically less than 10% whereas the open area in a packed bed is greater than 50%. Additionally, packing tends to have a lower liquid hold-up than trays. These characteristics combine to result in pressure drop for a theoretical stage being lower in packing than in trayed systems. This reduction in pressure drop can have a significant impact on tower and plant economics.

As with all situations, packed towers bring their own particular challenges. As an example, the same open area and reduced liquid hold-up that benefits pressure drop may result in lower liquid and vapour velocities, and therefore, an increased susceptibility of packed beds to accumulate solids in systems prone to polymerization or precipitation. Another common issue surrounds mechanical integrity. Packed bed towers are more susceptible to damage as a result of pressure surges as compared to trayed vessels. Finally, a critical consideration in the effective operation of a packed tower is the mechanism for fluid distribution. Poor vapour or liquid distribution can result in a significant efficiency reduction. In fact, the necessity of uniform distribution increases with the number of theoretical plates in a bed. This results in tower internal designers limiting bed length and redistributing process materials at specific and regular intervals.

With the increased use of packed beds, the application of radioisotope technology to troubleshoot problems has become popular. This paper, with the aid of case studies, describes the general principles of the radioisotope diagnostic technologies used in both trayed and packed bed towers that allow operating characteristics to be measured. It shows the benefits that can be gained by using the technologies as well as limitations and in some cases the necessary strategy of radioisotope technology integration to successfully determine the source of a particular mass transfer problem.

176a Tray Design Techniques at Low Liquid Load Conditions

Daniel R. Summers, Sulzer, 4019 South Jackson Avenue, Tulsa, OK 74107 and Andrew W. Sloley, VECO USA, Inc., 1313 Bay Street, Bellingham, WA 98225.

We (as an industry) have a difficult time describing minimum liquid load for many trayed distillation applications. For example, we prescribe two criteria at present for standard downcomers; minimum weir loading and spray (blowing) factor. For truncated downcomers there is a third term which has to deal with keeping the downcomer sealed against vapor by-pass. The problem is that the first two are ill defined and little is understood about them. In addition, the industry lacks the technical background as to why we even look at these values at all. Wash sections of towers, upper sections of low pressure distillation towers and separation of high boilers all can lead to tray designs that have very low liquid loads. Today many of these applications use random or structured packings as the preferred tower internal. However, as engineers we are to find the economical solution to a problem, and many times the economical solution may still be trays. This paper is intended to address this issue and help define a design approach with Picket Fence outlet weirs that can accommodate low liquid load tray design.

176b Distillation Column Flooding Predictor

M. Geoff Mphahlele and R. Bruce Eldridge. Chemical Engineering, University of Texas at Austin, 10100 Burnet Road, Room 1.310D, Austin, TX 78758

The Dzyacky distillation column flooding predictor is an advanced process control strategy that utilizes a patented (US 5,748,538) pattern recognition system to identify the onset of pre-flood conditions in distillation columns. By avoiding flooding, the column operates in a stable mode with high throughput and the process becomes more energy efficient.

To facilitate implementation of the predictor methodology, a general method for predicting the occurrence of flooding in liquid-gas mass transfer devices, specifically trayed distillation columns, is required. This research will firstly develop a generalized dynamic model to describe the transient behavior of a sieve-tray distillation column operation. This model will be based on predicting the dynamic response of liquid hold-up, liquid level, pressure drop and mass transfer efficiency. Existing state-of-the-art correlations will be used in the model.

The ultimate goal of the research is to predict a priori the pre-flood alarm values for column operation based on first-principles models (equilibrium and rate-based models) using column dimensions and other column specifications (e.g. feed rates, feed composition, reflux ratio etc). These alarm values are key to predicting a pre-flood condition. Currently, the plant operator determines these values by using rules of thumb or guesswork.

The dynamic model results will be validated using pilot plant data from the Separations Research Program at the University of Texas, Austin. This validation will also involve using the model to identify pre-flood patterns. The dynamic model results will also be compared to data obtained from a commercial ethane/ethylene distillation column. Wavelet-based methods will be applied to the data to remove the background noise prior to model validation

Keywords: Flooding prediction, Distillation, Equilibrium stage model, Model validation, Dynamic modeling and Simulation, Pattern recognition system, Wavelet method.

176c Computer-Aided Graphical Tool for Complex Column Design

Michaela Vrey, Simon Thornhill Holland, Diane Hildebrandt, and **David Glasser**. Center of Process and Materials Engineering (COMPS), University of the Witwatersrand, Private Bag 3, WITS 2050, Johannesburg, South Africa
Keywords: Column section, column profile maps, novel short-cut technique, coupled columns, computer-aided graphical tool

Abstract

Distillation is the most important separation process in the chemical and other related industries. Due to extensive theoretical and experimental research in the 20th century much is known about the distillation process in a one-feed two-product column. Based on this knowledge powerful rigorous process simulators such as ASPEN PLUS have been developed and are now available allowing the simulation of complex columns. For many people this is reason enough to declare the field of distillation as very well understood and therefore require no further research.

However complex columns, such as the divided wall column, are extremely difficult (if not impossible in a lot of cases) to design successfully using the process simulators available. This is due to the fact that these simulators give little insight into the behavior of the system and are generally not suitable for conceptual distillation process design. Therefore very few complex column designs exist in industry. Although complex columns seem difficult to operate, they may provide significant cost savings e.g. by substituting two or more simple columns by a single complex column. In other words the major problem in the development of new separation systems for industry is the lack of rapid and inexpensive screening and synthesis tools.

Until recently, no simple technique to model complex column configurations such as coupled columns (e.g. Petlyuk column), divided wall columns or columns with multiple feeds and product withdrawals was available.

We have developed a computer-aided graphical tool for complex column design. This approach is based on the works of Tapp [1] and Holland [2] who used the difference point equation and column profile maps (CPM's) to determine the composition profile in a column section. A column section is the generalized version of a stripping and/or rectifying section in a distillation column and is defined as a length of a column between points of addition or removal of material and/or energy. The advantage of this approach over the traditional rectifying and stripping profile method is that any column configuration can be modeled and entire maps (CPM's) can be generated as opposed to single profiles. As CPM's are transformed RCM's insight in finite reflux behaviour can be gleaned.

Using this technique we are currently able to design any complex column configuration and network and compare different configurations to determine the optimal one for a particular separation for ideal and selected non-ideal three component systems. This technique enables us to also determine feasible regions of complex column configurations. It provides insight into the dynamics of the Petlyuk column, which would enable the successful initialization of rigorous iterative design packages. We can show that the results obtained using our technique agree with a simulation done with ASPEN however only a fraction of the time is needed for the successful determination of the parameters.

The objective of this paper is to demonstrate this powerful technique by looking at a ternary separation problem using the divided wall column set-up. We will show that by having the product composition (top and bottom) and the feed composition we are able to determine i.e. the minimum reflux/reboil ratio, the range of feasible split ratios for the minimum reflux/reboil ratio, feed stage, side-draw stage and composition and the optimal split ratio for the minimum number of stages.

References

[1] Tapp, M., Holland, S.T., Hildebrandt, D., Glasser, D., "Column Profile Maps. 1. Derivation and Interpretation", Ind. Eng. Chem. Res., 2004, 43(2), pp364-374.

[2] Holland, S.T., Tapp, M., Hildebrandt, D., Glasser, D., Hausberger, B., "Novel Separation System Design Using Moving Triangles", Comp. & Chem. Eng., 2004, 29, pp 181-189.

176d Heat-Integrated Distillation Sequence Synthesis

Santosh Jain, Jin-Kuk Kim, and **Robin Smith**. Centre for Process Integration, The University of Manchester, Sackville Street, PO Box 88, M60 1QD, Manchester, United Kingdom

The presentation will highlight the design and synthesis of heat-integrated distillation sequence. The novel and systematic methodology has been developed to achieve the cost-effective and environmental-benign design of distillation systems, where the sequence, types and configurations of distillation columns are fully and simultaneously optimised with heat integration options.

The proposed methodology is able to screen, evaluate and optimise all the opportunities within the heat-integrated distillation system: i) distillation sequence, column type (simple vs complex), ii) many degrees of freedom in distillation (e.g. column pressure, feed condition, reflux ratio), iii) many heat integration options (e.g. Inter-column heat integration between condenser and reboiler, steam generation from condensers, feed preheating, etc), and iv) design interactions (e.g. operating conditions affect column temperatures and heat duties, interactions between two successive columns, etc).

The design framework is formulated as a mathematical optimisation problem that is solved using a genetic algorithm. Case studies will be provided to illustrate the significant cost savings achieved from the proposed approach.

176e Synthesis and Design of Energy Efficient Distillation

Angelo Lucia, University of Rhode Island, Dept of Chem. Engineering, Kingston, RI 02881, Ross Taylor, Clarkson University, Department of Chemical Engineering, Potsdam, NY 13699, and Amit Amale, Chemical Engineering, University of Rhode Island, 16 Greenhouse Road, Kingston, RI 02881-0805.

Batch and continuous distillation and crystallization have been the workhorses for separations in the petroleum, chemical, pharmaceutical, and other industries for many years, and this is unlikely to change. These unit operations, as well as others, will remain the primary means of separation in many industries for many years to come. However, distillation and crystallization consume significant amounts of energy even when compared to the US GNP. While some believe that distillation is a mature technology and that there is little to be gained from research in separations like distillation and crystallization, we disagree with this viewpoint for two reasons. First, with the recent significant increase in global energy demands and every indication that demand will remain high, it is important to consider ways of designing or retrofitting separations so they are energy efficient. Second, the approach taken in this work is a direct outgrowth of recent results that shed new light on residue curves and separation boundaries. It is unlikely that we would have uncovered the proposed characterization of energy efficient separations without our initial results.

Lucia and Taylor (2005) have recently presented a geometric methodology for finding exact boundaries in separation processes like azeotropic distillation, crystallization, and reactive distillation and show that for ternary mixtures all separation boundaries are given by locally longest residue curves that run from a given unstable node to all reachable stable nodes. For four-component mixtures, boundaries are local maxima in surface areas while for five or more components boundaries correspond to local maxima in volumes. This geometric theory has led to an efficient feasible path optimization algorithm for computing exact separation boundaries for batch or continuous separations. Moreover, rigorous proof and a number of challenging numerical examples have been used to validate the theory.

In this work, we conjecture that local maxima in line integrals, surface areas, and volumes (i.e., the longest paths) represent the most difficult and therefore the least energy efficient separations in a given separation region. Conversely, local minima in line integrals, surface areas, and volumes (i.e., the shortest paths) correspond to the easiest and most energy efficient separations. We quantify these conjectures and show that locally longest and shortest line integrals for residue curves do, in fact, correspond to the most and least difficult separations respectively. These facts provide definitive guidance for the synthesis, design, and retrofitting of energy efficient finite separators. Because we are interested in finite designs (i.e., finite stages and finite internal flows), distillation lines are used in place of residue curves. First, distillation lines at infinite reflux and infinite reboil are used to show that the minimum number of stages (or height of packing), Nmin, required to achieve a desired separation and therefore the easiest separation occurs in the neighborhood of the shortest distillation lines. Second, we show that the separations with the smallest minimum reflux, and thus the lowest energy demands, again appear in the neighborhood of the shortest distillation path. This approach allows us to easily screen alternative designs with regard to energy use and to find the most energy efficient separations for a variety of situations (i.e., fixed number of stages, fixed safety margins, batch operation, continuous operation, and so on). It also allows us to retrofit existing columns with a fixed number of stages to improve energy efficiency. In addition, our results are general and apply to crystallization, reactive separations, non-equilibrium models, and other separations. This work includes theoretical development as well as numerical experimentation and validation. Numerical examples and geometric illustrations are used to elucidate key concepts and to demonstrate the practical value of the proposed approach in finding energy efficient separator designs.

176f How to Decide When and How Much to Use Reactive Distillation: a Case Study on Mtbe System

Jean Lubilanji, David Glasser, and Diane Hildebrandt. Center of Process and Materials Engineering, University of the Witwatersrand, Private Bag 3, WITS 2050, Johannesburg, South Africa

Abstract

Process design deals with both synthesis and optimization steps. Here by process synthesis we mean, to create the best process structures, and by optimization we mean the choice of the best set of operating conditions. The intensive use of computer software has hugely increased our ability to find optimal solutions; thus finding the set of alternatives structures has become the bottleneck in the process design task.

In this paper we propose a simple method based on a thermodynamic approach to decide on the best process structure. We consider the reactive distillation process as a combination of the individual separation and reaction vector fields and then look at how these vectors combine in different regions of the composition space. These geometric insights are then interpreted in term of their structural implications. These vector maps (together with the Iso-separation and Iso-rate contours) provide a powerful tool at the early stage of process synthesis when one is making decisions about the structure of the plant. We illustrate the ideas by looking at the MTBE system.

Author Keyword: Reactive distillation References

1. Ung, S., Doherty.M.F," Vapor-liquid phase equilibrium in systems with chemical reactions" Chemical Engineering Science, 1995, 50, pp.23-48 2. Barbosa, D., Doherty.M.F,"The simple distillation of homogeneous reacting mixtures" Chemical Engineering Science, 1988, 43, pp.541-550 3. Hauan, S., Westerberg, A., Kristian, L.,"Phenomena-based analysis of fixed points in reactive separation systems" Chemical Engineering Science, 2000, 55, pp. 1053-1075 4. Hauan, S. and K.M.Lien,"A phenomena based design approach to reactive distillation" Trans.I.Chem.E, 1998, 76(A), pp.396-407 5. Hoffmaster, W., Hauan, S.," Difference points in reactive and extractive cascades. III-Properties of column section profiles with

arbitrary reaction distribution" Chemical Engineering Science, 2004, 59, pp. 3671-3693 6. Huang, Y-S., Qi, Z., Schlunder, E-U., Sundmacher, K," Residue curve maps for reactive membrane separation" Chemical Engineering Science, 2004, 59 (14), pp.2863-2879

American Institute of Chemical Engineers
3 Park Avenue, New York, NY
10016-5991

ISBN: 978-1-60423-527-2